Self-Assessment Color Review of

# Clinical Anatomy

## Edward J Evans
MB BS (Lond) LRCP MRCS MSc (Surrey)
Lecturer
Anatomy Unit, Cardiff School of Biosciences,
Cardiff University, UK

## Bernard J Moxham
BSc BDS PhD(Bristol)
Professor and Head of Anatomy
Anatomy Unit, Cardiff School of Biosciences,
Cardiff University, UK

## Richard L M Newell
BSc MB BS (Lond) FRCS (Eng)
Clinical Anatomist
Anatomy Unit, Cardiff School of Biosciences,
Cardiff University, UK

## Robert M Santer
BSc PhD (St Andrews)
Senior Lecturer
Anatomy Unit, Cardiff School of Biosciences,
Cardiff University, UK

LIPPINCOTT WILLIAMS & WILKINS
A Wolters Kluwer Company
Philadelphia · Baltimore · New York · London
Buenos Aires · Hong Kong · Sydney · Tokyo

First published in the United States of America in 1999 by
Lippincott–Williams & Wilkins, 227 East Washington Square
Philadelphia, PA 19106
ISBN: 07817–1905–4

A CIP catalogue record for this book is available from the British Library.

For full details of all Manson Publishing titles, please write to:
Manson Publishing Ltd, 73 Corringham Road, London  NW11 7DL, UK.

Design and layout: Judith Campbell of The Little Round Consultancy.
Colour reproduction: Jade Reprographics, Braintree, UK.
Printed by Grafos SA, Barcelona, Spain.

# Preface

This book has been written and designed to appeal to medical students and to post-registration trainees. It allows both types of reader to cover the whole field of topographical (gross) anatomy, whether in the context of regional, systematic, or clinical anatomy. Indeed, we have conceived the book from the conviction that, without a basic understanding of anatomy, the medical practitioner will remain ill-equipped.

All of the scientific and clinical disciplines, in many medical schools throughout the world, are concurrently involved in major educational revisions. Increasingly, both undergraduate and postgraduate training schemes are becoming more 'integrated' in order to emphasize clinical relevance. Furthermore, teachers are expected to establish, and define, 'core' material so as to lighten the demands placed upon novice learners! However the new courses and training programmes are organized, anatomy remains a core subject, provided that its relevance to clinical practice is highlighted. It is for this reason that most of the questions in this self-assessment book have the format of case histories or clinical puzzles that require anatomical information for their elucidation. These latter should not only confirm the importance of basic anatomy to the clinical situation and aid motivation but also conform with modern teaching practices that encourage problem solving.

An important component of the new courses is 'student-centred learning'. This has the aim of trying to make students more responsible for their own education, and develop an independence of mind that will fit them for continuing education once their formal courses are completed. For student-centred learning to be effective, it is important that students should assess their progress by frequent testing of their knowledge and skills. This book has been written to help students at various stages of their courses to revise and to assess their progress; it is not purely for the purpose of passing examinations.

With respect to the content of this book, we are only too aware of the pressures nowadays placed upon students undertaking medical and surgical training programmes. Whilst it is undoubtedly tempting to cut material from courses, it is exceedingly difficult to do this without reducing standards. This book was written with the aim of presenting a body of material that we believe approximates the minimum standards required for those undertaking courses that include topographical anatomy. It should enable students to test themselves on a wide range of essential topographical material and will point them to those areas that need further study.

It is also our belief that a proper appreciation of anatomy does not rely upon the assimilation of a mass of facts. Anatomy is essentially a 'visual' subject (including a three-dimensional appreciation of human structure) and cannot be mastered simply by reading a text. In addition to some true/false questions to test factual knowledge, we have therefore provided the opportunity for readers to assess their development of a visual appreciation of important anatomical features by incorporating numerous full-colour illustrations. These illustrations do not merely cover human dissection (of a quality typically to be found in any dissection room) but also include radiographs, magnetic resonance images (MRIs), computed tomograph (CT) images, and clinical photographs.

# Acknowledgements

A book of this kind, which brings together preclinical science and clinical case histories, could not have been compiled without the generous support of many friends and colleagues (from a variety of medical disciplines), who loaned material and photographs. We would therefore like to acknowledge and thank Mr J. Stansbie (Walsgrove Hospital, Coventry), Mr E. Roberts, Professor G. Roberts, Dr A. Williams, Dr A. Fraser, Dr J. Potts, Dr M. Crane, Professor R.E. Mansel, Mr R.D. Weeks, Professor M.F. Scanlon, Mrs L. Beck, Mr M.H. Wheeler, Mrs C. Lane, Dr M.D. Hourihan, and the Department of Medical Illustration (all at the University of Wales, College of Medicine).

Within the Anatomy Unit at Cardiff, we would like to say an especial thank you to our demonstrators and prosectors and to our students who gave advice freely and who sometimes acted as 'guinea pigs' to test the questions. A major debt is also owed to our secretaries, Miss A. Filice and Mrs C. Malpass, for their clerical assistance and for keeping the office happy! Furthermore, we could not have produced this book if it was not for the skill and enthusiasm of our photographer, Mr G. Pitt.

Finally, we would wish to remember all those who bequeath their bodies for anatomical examination and who, by such kindness, do so much for medical education and research.

# THE LOCOMOTOR SYSTEM

## Upper Limb and Back

For each of questions 1–20, state whether the statement provided is true or false.

1    Abduction of the shoulder takes the upper arm further away from the body.

2    The scapula has no direct bony attachment to the axial skeleton.

3    The coraco-acromial ligament forms part of the capsule of the shoulder joint.

4    The shoulder joint is partly innervated by the suprascapular nerve.

5    Subscapularis is attached to the dorsal surface of the scapula.

6    Teres major is one of the rotator cuff muscles.

7    Supraspinatus is attached to the greater tubercle of the humerus.

8    The median nerve is derived from posterior divisions of the brachial plexus.

9    The upper trunk of the brachial plexus contributes to the ulnar nerve.

10 Biceps brachii is attached to the anterior surface of the humerus.

# 1–10: Answers

1    TRUE – abduction is defined as a movement away from the sagittal plane.

2    TRUE – the scapula articulates with the clavicle, and the clavicle with the axial skeleton.

3    FALSE – it bridges between the coracoid process and acromion of the scapula, but plays no part in the structure of the shoulder joint.

4    TRUE – Hilton's law says that the nerve supplying a muscle also supplies the joint that the muscle moves.

5    FALSE – it is attached to the deep or ventral surface.

6    FALSE – teres minor is the cuff muscle.

7    TRUE – its tendon blends with the capsule of the shoulder and the adjacent tendon of infraspinatus.

8    FALSE – the median nerve comes from anterior divisions and will supply flexor muscles (among others).

9    FALSE – the ulnar nerve gets its fibres only from the lower trunk.

10    FALSE – biceps is not attached to the humerus at all; it is attached to the scapula and the radius, and (indirectly via the bicipital aponeurosis) to the ulna.

11  Brachialis is an important flexor of the elbow.

12  Triceps brachii is attached to the olecranon of the ulna.

13  Triceps brachii is innervated by the radial nerve.

14  All the extensor muscles in the upper limb are innervated by the radial nerve.

15  Flexor digitorum superficialis is innervated by the median nerve.

16  The inferior radio-ulnar joint lies immediately dorsal to pronator teres.

17  The scaphoid bone articulates with the lower end of the radius.

18  The tendons of flexor digitorum profundus to the fingers share fibrous sheaths with the superficialis tendons in each finger.

19  The interossei of the hand are innervated by the ulnar nerve.

20  The abductor pollicis brevis is innervated by the median nerve.

# 11–20: Answers

11  TRUE – it is at least as important as biceps.

12  TRUE – this is the insertion of triceps.

13  TRUE – the radial nerve supplies all the extensor muscles of the upper limb.

14  TRUE – (see 13) but the radial nerve does not *only* supply extensors.

15  TRUE – the median nerve is attached to the deep (dorsal) surface of this muscle.

16  FALSE – it lies dorsal to pronator quadratus.

17  TRUE – the scaphoid and the lunate are both involved in the wrist joint. In full abduction of the wrist, the lunate is separated from the lower end of the ulna by a triangular fibrocartilage.

18  TRUE – the two tendons run in a common fibrous sheath along the palmar sur face of the finger. They are also invaginated into a synovial sheath.

19  TRUE – wasting of the first dorsal interosseus muscle is an important sign of ulnar nerve damage.

20  TRUE – wasting of the abductor pollicis brevis is an important sign of median nerve damage.

**21** Identify the labelled features (**A–P**) on the photograph (**21**) of the upper limb (anterior view of left arm and posterior view of right arm).

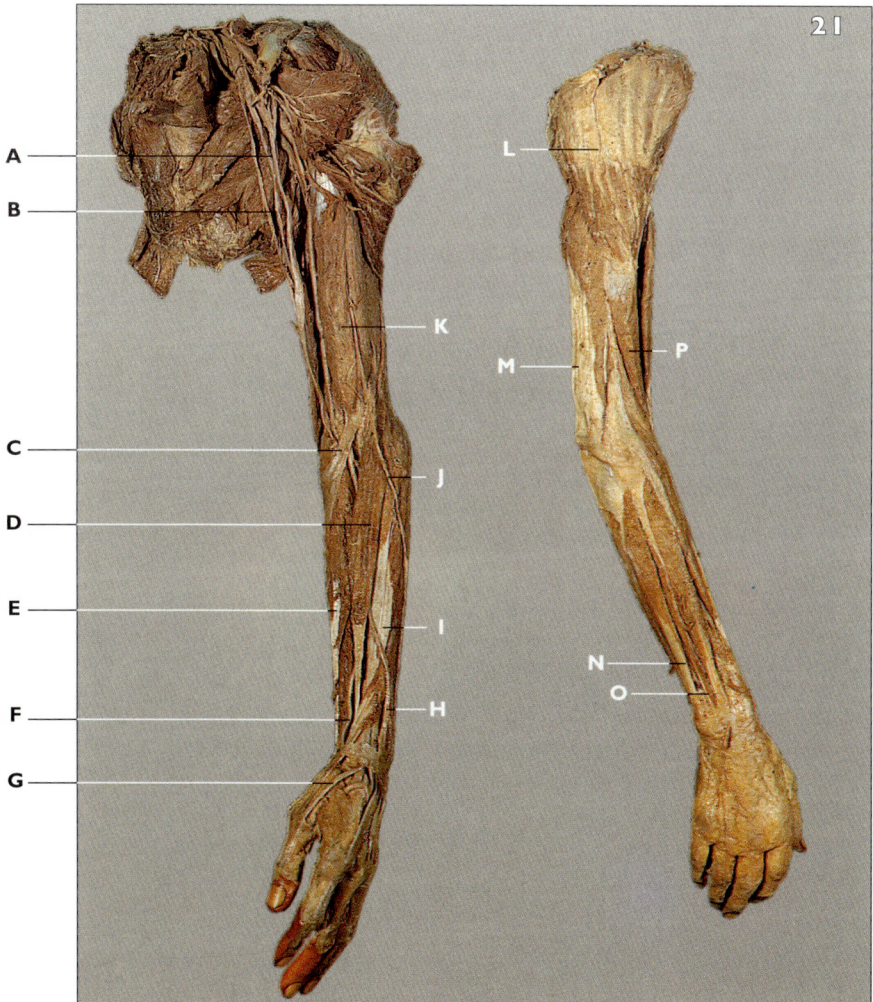

21   A: median nerve (note that as a result of dissection, the nerve is no longer in contact with the brachial artery). B: ulnar nerve. C: bicipital aponeurosis. The brachial artery is clearly seen in its normal position beneath the aponeurosis. D: brachioradialis. E: flexor carpi radialis. Distinguishable because of its lateral and superficial position. F: radial artery. Note that in this individual it has an anomalous course; it is just visible passing normally across the floor of the 'anatomical snuffbox'; it then passes deep to the insertion of extensor carpi radialis longus, which is unusual; finally, it is easily seen passing between the two heads of the first dorsal interosseous, as is usual. G: extensor pollicis longus tendon, forming one of the boundaries of the 'anatomical snuffbox'. H: superficial terminal branch of the radial nerve, supplying a variable amount of skin on the lateral part of the back of the hand. I: extensor carpi radialis longus. Longus is more superficial than brevis, so is visible in this view. J: lateral cutaneous nerve of the forearm, the terminal part of the musculocutaneous nerve. K: biceps brachii. L: deltoid. M: triceps. N: extensor carpi ulnaris. O: extensor digitorum. P: brachialis. The edge of the muscle shows clearly between biceps and triceps brachii.

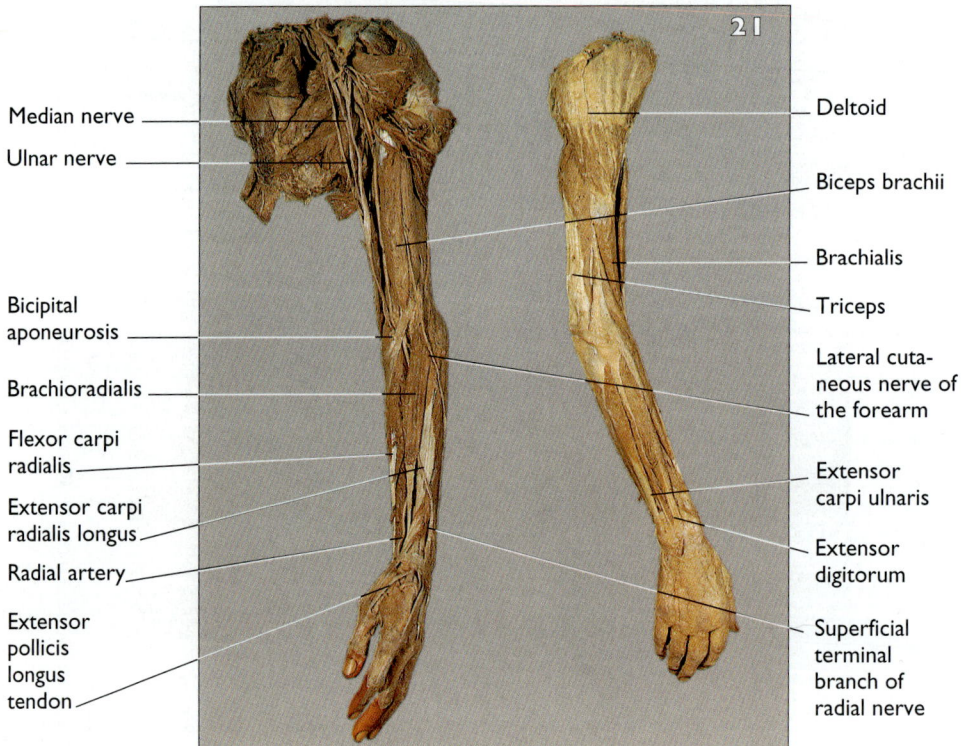

Median nerve

Ulnar nerve

Bicipital aponeurosis

Brachioradialis

Flexor carpi radialis

Extensor carpi radialis longus

Radial artery

Extensor pollicis longus tendon

Deltoid

Biceps brachii

Brachialis

Triceps

Lateral cutaneous nerve of the forearm

Extensor carpi ulnaris

Extensor digitorum

Superficial terminal branch of radial nerve

22   Dissection of the hand (**22**). Identify the features labelled **A–G**.

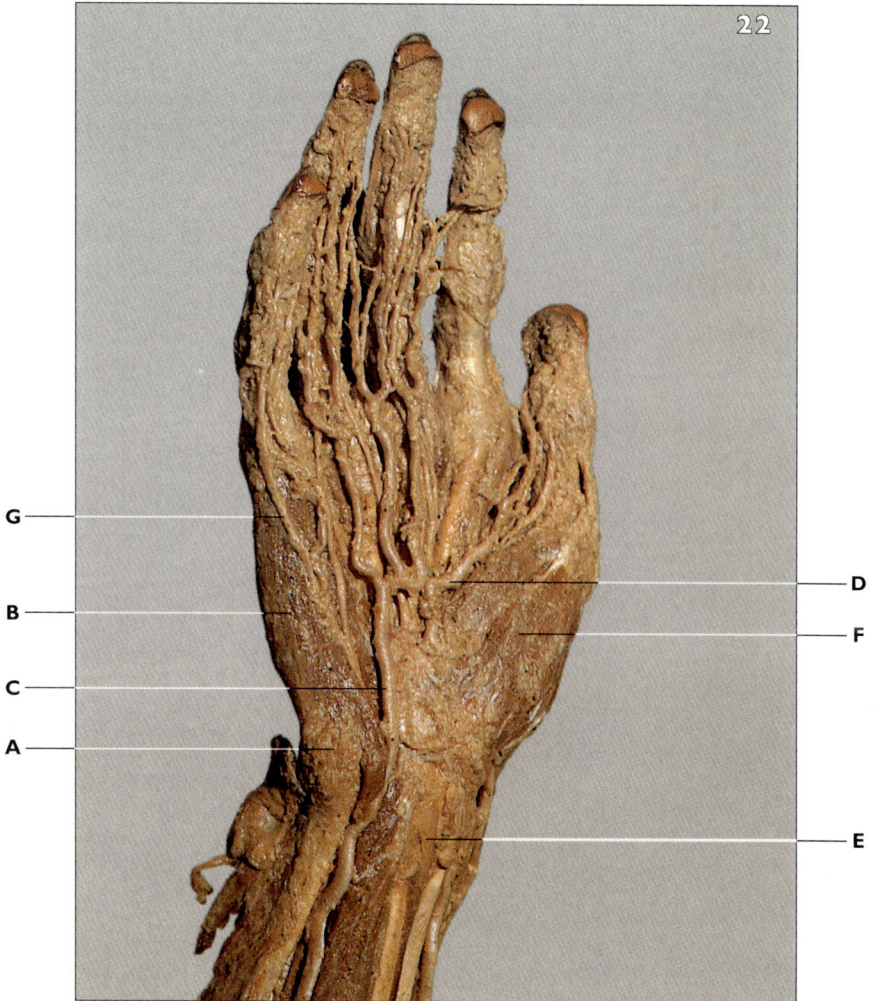

# 22: Answers

**22**  A: pisiform bone. B: abductor digiti minimi. C: ulnar artery. D: superficial palmar arch. In this specimen, the arch is large and supplies the thumb, but its connection with the radial artery is invisible. E: median nerve, wrapping around the lateral border of flexor digitorum superficialis and disappearing under the flexor retinaculum. F: abductor pollicis brevis. It is often hard to separate the muscles of the thenar eminence, but abductor pollicis brevis always occupies the centre of the muscle mass when seen from the front. G: digital branch of the ulnar nerve.

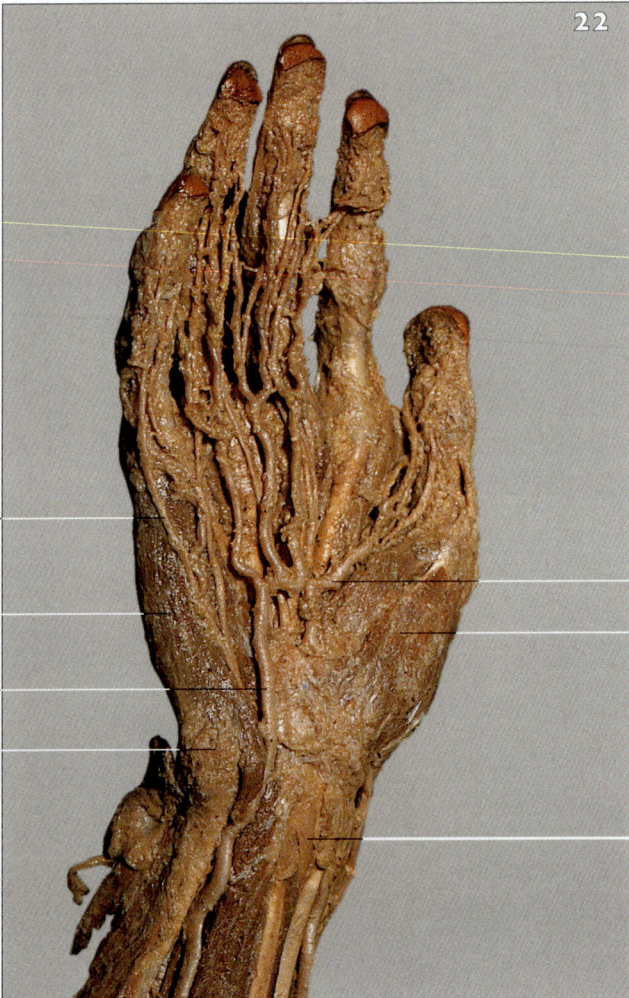

Digital branch of the ulnar nerve

Abductor digiti minimi

Ulnar artery

Pisiform bone

Superficial palmar arch

Abductor pollicis brevis

Median nerve

**23** A 55-year-old woman noticed that the skin of her right breast appeared to become stuck to a painless lump in the breast (**23a**). She consulted her GP who, as part of his clinical examination, confirmed the 'tethering' of the lump to the skin and looked for 'tethering' to deeper structures. He also examined the lymph nodes to which tissues in the area of the lump would drain.

**i.** Which structures lie deep to the breast that may be invaded by a malignant tumour to cause 'tethering' of the lump?

**ii.** To which lymph nodes do the tissues of the breast normally drain?

**iii.** To which lymph nodes might this patient's tumour drain?

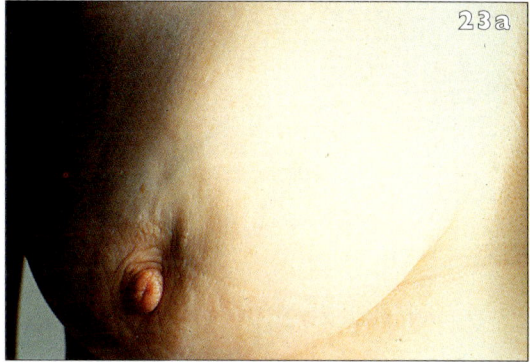

**24** A 25-year-old man injured his shoulder while playing football. Clinical and radiological examination (**24**) indicated a dislocation.

**i.** Which nerve runs around the surgical neck of the humerus and is at risk of damage both by dislocation of the shoulder and by attempts at reduction?

**ii.** Which muscles does the nerve supply?

**iii.** Does the nerve supply any skin that the doctor could use to test for the integrity of the nerve before (and after!) reduction of the dislocation?

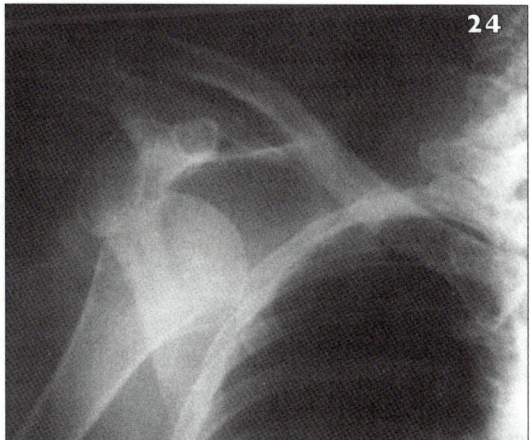

**23** **i.** Pectoralis major and the deep fascia over it.
**ii.** Mainly to the pectoral group and then to the central and apical groups of axillary nodes (**23b**). Some lymph passes directly to the apical groups. Some lymph from the medial half of the breast drains to the parasternal lymph nodes along the internal thoracic artery.
**iii.** To any of the nodes listed above.

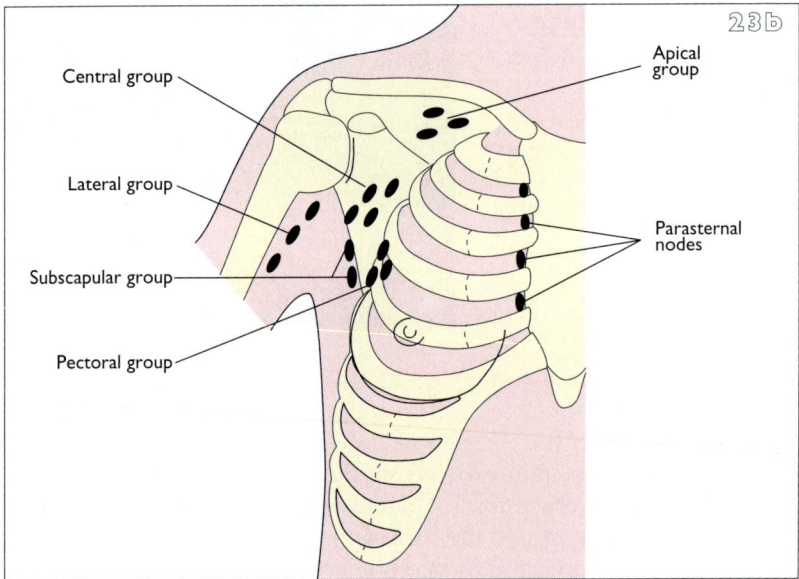

**24** **i.** Axillary.
**ii.** Deltoid and teres minor.
**iii.** Yes, over the insertion of deltoid on the upper outer part of the arm.

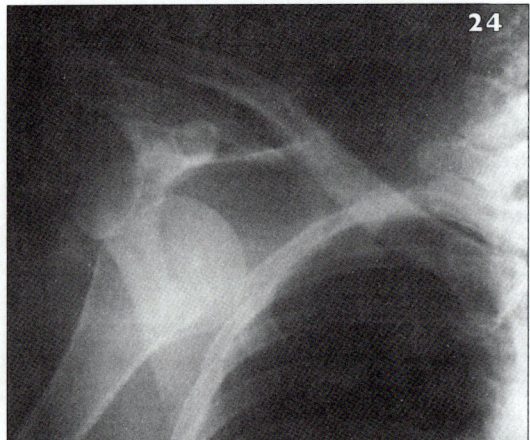

25 A 55-year-old woman was knocked down by a car and injured her right arm. The radiograph (25) showed a fracture to the middle of the shaft of the humerus.

i. Which nerve might be damaged by this fracture?

ii. Which groups of muscles does the nerve supply below the level of the mid-shaft of the humerus?

iii. Does the nerve have any cutaneous supply below this level?

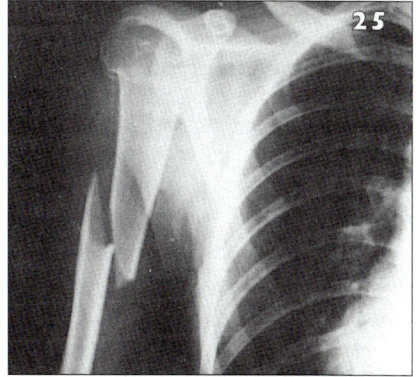

26 A middle-aged man is brought to the accident and emergency department. He is bleeding heavily from a wound in the upper thigh and his trousers are completely soaked with blood. He is semiconscious and very restless. The casualty officer wishes to start a blood transfusion.

i. Although there are superficial veins in the cubital fossa, the casualty officer does not use them in this case. Why not?

ii. At which other site(s) in the upper limb are superficial veins commonly present and easy to cannulate through the skin for a transfusion?

iii. Where else in the upper limb is a vein constantly present that is suitable for a 'cut-down' in an emergency?

27 A 40-year-old man caught his arm in farm machinery, sustaining a twisting injury of the whole of the right upper limb. The major injury was a compound fracture of both bones of the forearm (27).

i. Name the muscles through which the exposed bones must have passed.

ii. Which nerve supplying these muscles may also have been damaged?

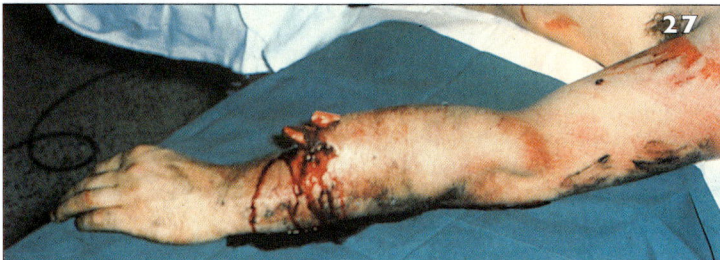

**25  i.** The radial nerve.
**ii.** All extensors of the wrist, thumb, and fingers; brachioradialis; supinator; and abductor pollicis longus. The radial nerve also supplies anconeus. Paralysis of the extensors of the wrist causes a condition known as wrist drop.
**iii.** Yes. It supplies an area of skin on the lateral part of the back of the hand. The posterior cutaneous nerve of the forearm is usually given off higher than midshaft.
Note that there is also a fracture of the surgical neck of the humerus in this patient.

**26  i.** First, because these peripheral veins are likely to be empty and therefore impalpable in a shocked patient; second, because the patient is likely to bend his elbow and dislodge the cannula.
**ii.** On the back of the hand, and the back or lateral side of the lower end of the radius (see arrows, **26a**).
**iii.** In the groove between pectoralis major and deltoid (cephalic vein). Note (see **26b**) that although the cephalic vein is not visible, the delto-pectoral groove is easily seen (arrows).

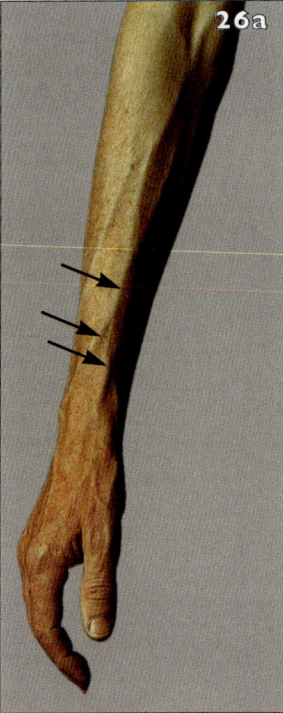

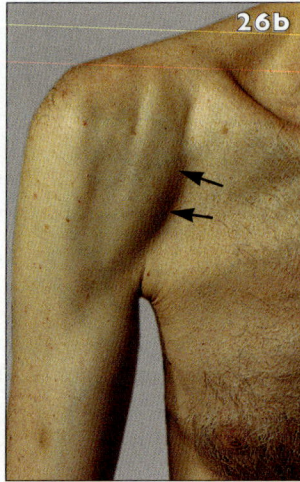

**27  i.** Brachioradialis, extensors carpi radialis longus and brevis, extensor digitorum communis, extensor carpi ulnaris, abductor pollicis longus, extensor pollicis longus.
**ii.** Posterior interosseous (deep) branch of the radial nerve.

28 A 28-year-old builder has cut the front of his wrist on broken glass. Before suturing the wound, the casualty officer wishes to check for injury to tendons crossing the wrist.
i. Which are the important tendons that cross the front of the wrist superficially?
ii. Which movements would you test to check for integrity of these tendons?
iii. Which important blood vessels cross the front of the wrist?

29 A 50-year-old man was injured by flying glass in an explosion. He sustained multiple lacerations of his left arm and forearm. After 3 months the lacerations had healed but he still complained of numbness of his little finger and the medial side of his hand, and difficulty in fully straightening his ring and little fingers. On examination, his GP noticed muscle wasting, especially between the first and second metacarpals.
i. Which muscles can be seen to be wasted in the illustration (29)?
ii. Which nerve has been injured?
iii. Which other muscles does this nerve supply in the hand?
iv. What is the name given to the typical deformity of the hand caused by injury to this nerve at the wrist?

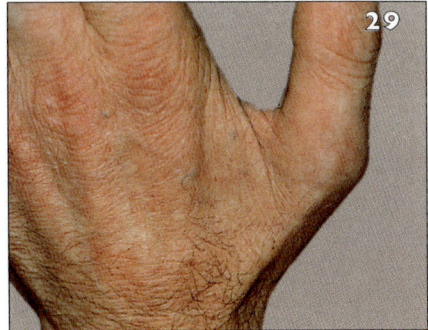

**28** The illustration (28) shows dissections of the tendons of the front of the wrist.

**i.** Flexor digitorum superficialis, flexor carpi radialis, flexor carpi ulnaris. (Palmaris longus can hardly be called important, flexor digitorum profundus is not superficial.)

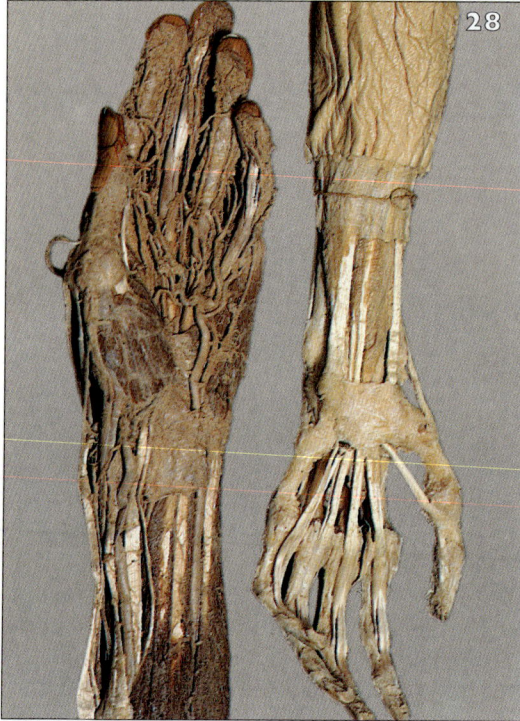

**ii.** Flexion of the wrist to test the flexors carpi. When testing flexor digitorum superficialis you must not be confused by flexor digitorum profundus, which also flexes the fingers. Ask the patient to flex each finger at the proximal interphalageal joint while the other fingers are held in extension. If the distal interphalangeal joint does not flex during this manoeuvre, flexor digitorum superficialis is intact.

**iii.** The radial and ulnar arteries.

**29  i.** Wasting of the first dorsal interosseus, between the first and second metacarpals, is obvious. There are also hollows between the other metacarpals, indicating wasting of the remaining dorsal interossei.

**ii.** The ulnar nerve. It supplies the dorsal interossei and the skin of the little finger (amongst others), so this must be the damaged nerve.

**iii.** All the interossei; the medial two lumbricals, the flexor, abductor, and opponens digiti minimi; and the adductor pollicis. The unlar nerve often shares the supply of flexor pollicis brevis with the median nerve.

**iv.** Claw hand. Because of the paralysis of the intrinsic muscles the patient is unable properly to flex the metacarpophalangeal (MP) joints or to extend the interphalangeal (IP) joints of the affected fingers. The unrestrained action of the long flexors and extensor is to flex the IP joints and extend the MP joints.

30 A 20-year-old man fell off his bicycle and landed on his outstretched hand. On examination, all movements of his wrist were painful and there was tenderness on pressure in the 'anatomical snuffbox'. The radiograph (30) taken on the day of injury appeared normal.

i. Where is the anatomical snuffbox and what are its boundaries?

ii. Which bones lie in the floor of the snuffbox?

iii. The casualty officer decided to apply a plaster, even though no fracture was seen on the radiograph. Which bone did he suspect might be damaged?

iv. Which other structure crosses the floor of the snuffbox?

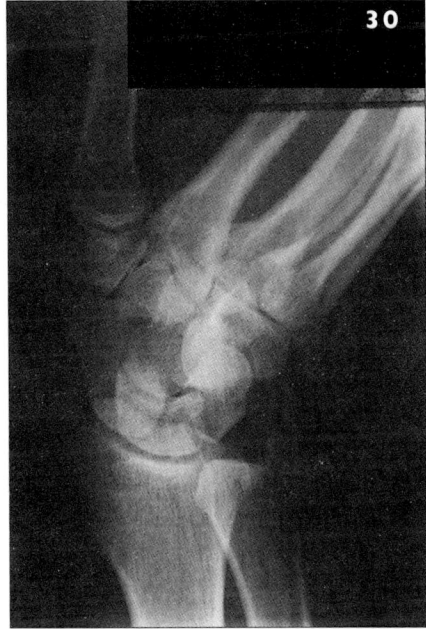

31 A patient attends the accident and emergency department with a laceration of the middle finger. The doctor wishes to anaesthetize the nerves to the finger by injecting local anaesthetic before suturing the laceration.

i. Which nerves supply the middle finger?

ii. Where, exactly, do these nerves run?

iii. Where is their cutaneous distribution?

iv. What other structures run with the nerves?

v. Why does the doctor check that the local anaesthetic preparation he is using does not contain adrenaline?

32 i. Where at the elbow would you expect to feel the brachial artery pulsating?

ii. Where in the fingers do the digital nerves and arteries run?

iii. Where does the ulnar artery cross the wrist?

iv. What is the normal relationship of the olecranon and the two epicondyles on examination of the elbow?

**30 i.** On the lateral side of the wrist, between the tendons of extensor pollicis longus and brevis. Abductor pollicis longus lies beside extensor pollicis brevis and is sometimes a larger tendon, but is on the side away from the snuffbox.
**ii.** Scaphoid and trapezium.
**iii.** Scaphoid. It is notoriously difficult to see a recent fracture of this bone on a radiograph.
**iv.** The radial artery. It may often be felt pulsating here.

**31 i.** Two palmar digital branches of the median nerve and two dorsal digital branches of the radial nerve.
**ii.** The palmar digital branches run along the sides of the fingers, approximately along a line joining the tips of the flexor creases of each joint. They are thus close to the palmar surface at the metacarpophalangeal joint, but a little more dorsal from the proximal interphalangeal joint onwards. The dorsal digital branches also run along the sides of the fingers, but near the dorsal surfaces.
**iii.** Dorsal digital branches in the middle finger supply the dorsal surface over the proximal phalanx. The palmar digital branches supply the remainder of the skin. In the index finger, the dorsal digital branches supply the skin over the dorsum of both proximal and middle phalanges. In all cases, there is considerable overlap (31).
**iv.** Digital arteries and veins.
**v.** Adrenaline causes constriction of blood vessels. It is sometimes used to reduce the rate at which anaesthetic is carried away from a site. In the finger, there is a risk that constriction of both digital arteries would completely cut off the blood supply to the finger, causing permanent damage.

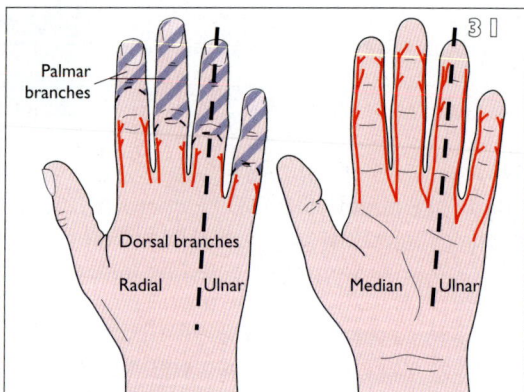

**32 i.** In the midline of the anterior surface. It is most easily felt just deep to the medial edge of the bicipital aponeurosis.
**ii.** Along the sides of the fingers. They lie deep to a line joining the tips of the flexor creases at the metacarpophalangeal and interphalangeal joints.
**iii.** Immediately lateral to the pisiform bone. It is close to the ulnar nerve here.
**iv.** They form a triangle when seen from behind with the elbow flexed to a right angle. The olecranon lies distal to the line joining the epicondyles.

## Lower Limb

For each of questions 33–52, show whether the statement provided is true or false.

33  Lymph from the perineum drains to the inguinal nodes.

34  The acetabulum of the hip bone is entirely lined with articular cartilage.

35  The iliofemoral ligament is important in limiting flexion of the hip joint.

36  The saphenous nerve runs in the subsartorial canal for part of its course.

37  The posterior cruciate ligament is attached to the anterior part of the tibial plateau.

38  The anterior cruciate ligament of the knee resists forward movement of the tibia on the femur.

39  The medial collateral ligament of the knee blends with the joint capsule.

40  The lateral collateral ligament of the knee is not attached to the tibia.

41  Rectus femoris is part of the quadriceps femoris.

42  Inversion and eversion occur at the subtalar joint.

# 33–42: Answers

33  TRUE – if these nodes are abnormal, not only the lower limb but also the perineum and lower abdominal wall should be examined.

34  FALSE – the cartilaginous lining is horseshoe shaped, leaving an area inferiorly that is devoid of cartilage.

35  FALSE – it is important in limiting extension.

36  TRUE – it then emerges from beneath the medial border of sartorius and continues   down the leg with the great saphenous vein.

37  FALSE – the cruciates are named after the positions of their attachments on the tibia.

38  TRUE – it is very important in resisting the gliding movement that would other wise occur. It also resists rotation of the tibia.

39  TRUE – it is closely blended with the capsule, unlike the lateral collateral ligament.

40  TRUE – it is attached to the head of the fibula.

41  TRUE – the parts of quadriceps femoris are rectus femoris, vastus lateralis, vastus intermedius, and vastus medialis.

42  TRUE – this, together with the talonavicular, is the main joint at which inver-sion   and eversion occur. The calcaneocuboid joint is also involved.

43 Tibialis anterior can act as a plantar flexor of the ankle when the Achilles tendon is damaged.

44 The hamstring muscles flex both the knee and the hip.

45 The adductor longus is attached close to the pubic tubercle proximally.

46 Damage to the obturator nerve will cause weakness or paralysis of the adductor muscles.

47 The obturator nerve supplies no skin

48 The extensor hallucis can help to dorsiflex the ankle.

49 The anterior tibial artery runs in the anterior osteofascial compartment of the leg.

50 In the stance phase of gait, lower limb muscles mainly act 'in reverse', from insertion to origin.

51 The great saphenous vein is connected to the deep venous system by veins that perforate the deep fascia.

52 The profunda femoris artery is incapable of anastomosis with branches of the popliteal artery.

43 FALSE – tibialis anterior is a dorsiflexor (and invertor) of the ankle.

44 FALSE – they flex the knee but extend the hip.

45 TRUE – its tendon makes a useful landmark for locating the tubercle, which is of importance in distinguishing different types of inguinal hernia.

46 TRUE – they are supplied by the obturator nerve.

47 FALSE – it supplies a small area of skin on the medial side of the thigh above the knee.

48 TRUE – there are no muscles whose *only* action is dorsiflexion of the ankle, so the extensors of the toes are needed to help perform the action. Try dorsiflexing your ankle against resistance, first with toes fully flexed, then with them extended.

49 TRUE – here it and the deep peroneal nerve are at risk in the 'compartment syndrome' involving this compartment.

50 TRUE – this important fact is often forgotten. Lower limb muscles are important dynamic stabilizers of ankle, knee, and hip when acting in a distoproximal manner.

51 TRUE – these perforating veins have valves that prevent flow from deep to superficial. Incompetence of these valves contributes to the development of varicose veins.

52 FALSE – the lowest perforating branch of the profunda femoris artery anastomoses with the ascending muscular branches of the popliteal, providing an important collateral circulation if the femoral artery is obstructed in the adductor canal.

53  Identify the labelled features (**A–R**) on the photographs (**53a** anterior and **53b** posterior) of the right lower limb.

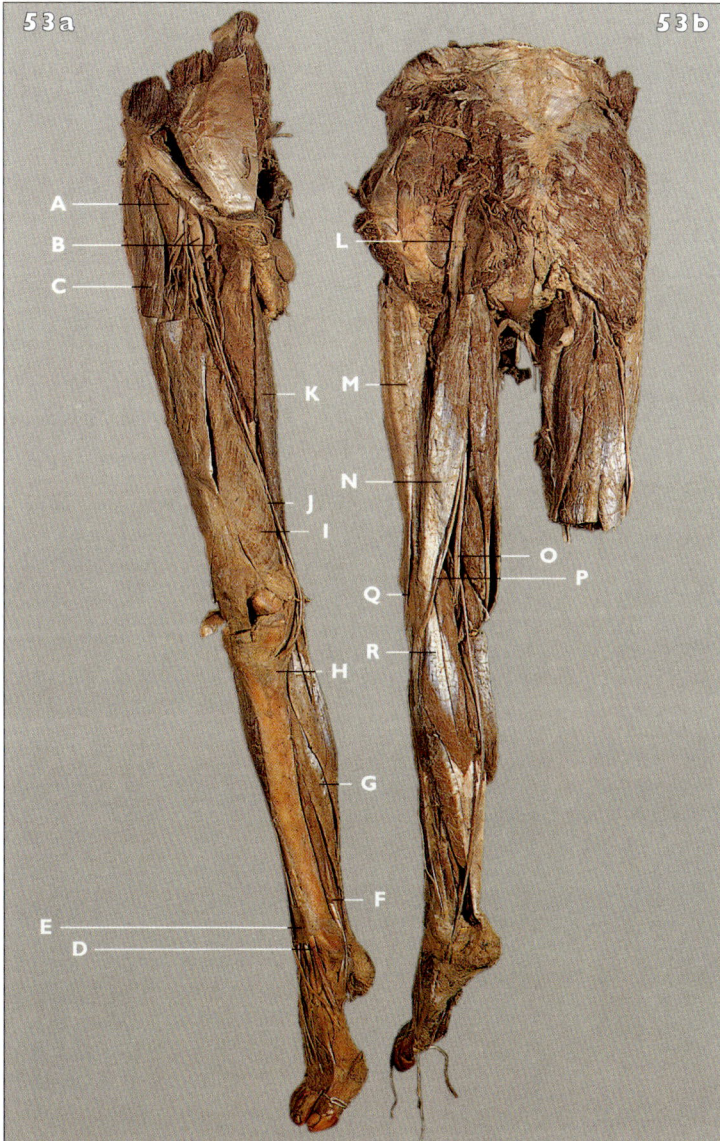

53 **A:** iliacus. **B:** femoral nerve artery and vein; nerve is most lateral, vein most medial. **C:** rectus femoris (cut and the upper part partly reflected). **D:** tibialis anterior. **E:** extensor retinaculum. **F:** flexor hallucis longus. **G:** soleus. **H:** the 'pes anserinus' or 'goose's foot'; this is the name sometimes given to the combined insertions of sartorius, gracilis and semitendinosus; sartorius has been removed in this dissection (note that there are at least two other structures termed 'pes anserinus' in human anatomy – the filaments of the facial nerve emerging from the parotid gland and the plexus of the infraorbital nerve ramifying on the cheek). **I:** vastus medialis. **J:** the saphenous nerve; this nerve passes deep to sartorius, but sartorius has been removed in this dissection, revealing the nerve in the subsartorial canal. **K:** gracilis; this could be confused with adductor magnus because the angle at which the muscle is seen makes it appear thicker than it really is. **L:** sciatic nerve. **M:** vastus lateralis, seen through the lateral intermuscular septum and the lateral fascia lata. **N:** biceps femoris. **O:** tibial nerve. **P:** common peroneal nerve. **Q:** iliotibial tract inserting into the lateral condyle of the tibia. **R:** lateral head of gastrocnemius.

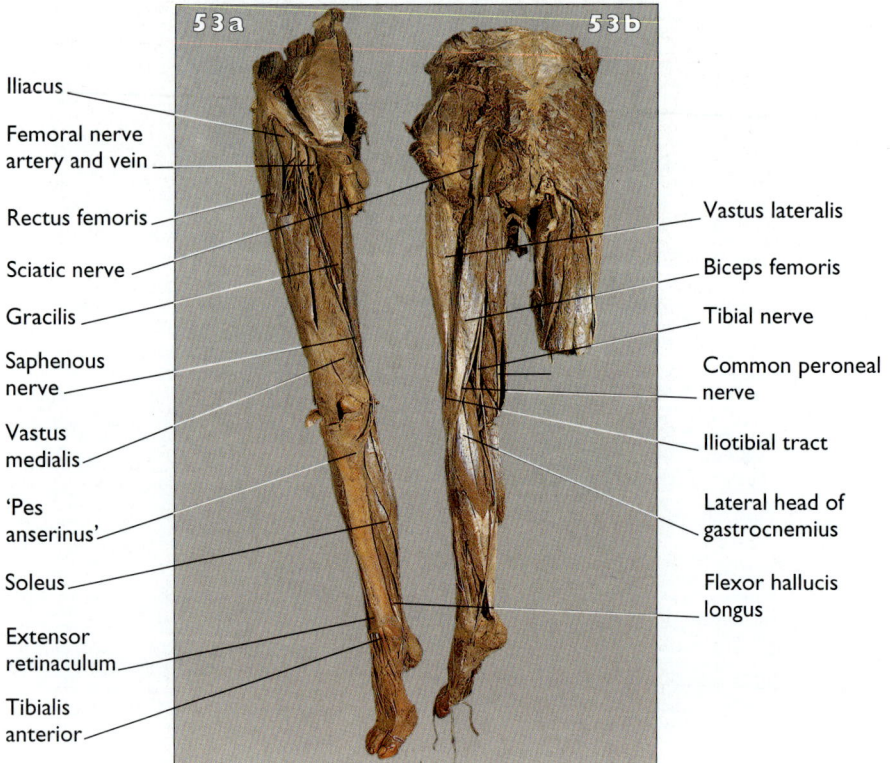

**54** Below is shown an axial (transverse) section through the lower leg (54), just below the metaphysis of the tibia. Identify the features labelled A–K.

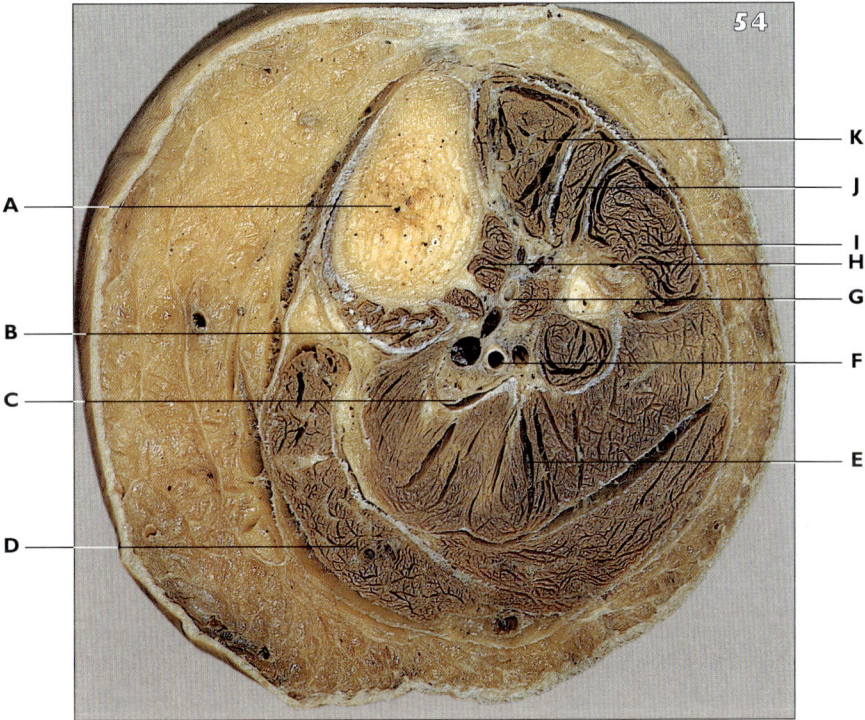

# 54: Answers

**54** **A:** tibia. **B:** popliteus. **C:** tibial nerve. **D:** gastrocnemius (medial head). **E:** soleus. **F:** posterior tibial artery. **G:** anterior tibial artery; note that, at this level, the artery has not yet passed anterior to the interosseous membrane. **H:** tibialis posterior. **I:** peroneal muscles; probably only peroneus longus is present at this level. **J:** extensor digitorum longus. **K:** tibialis anterior.

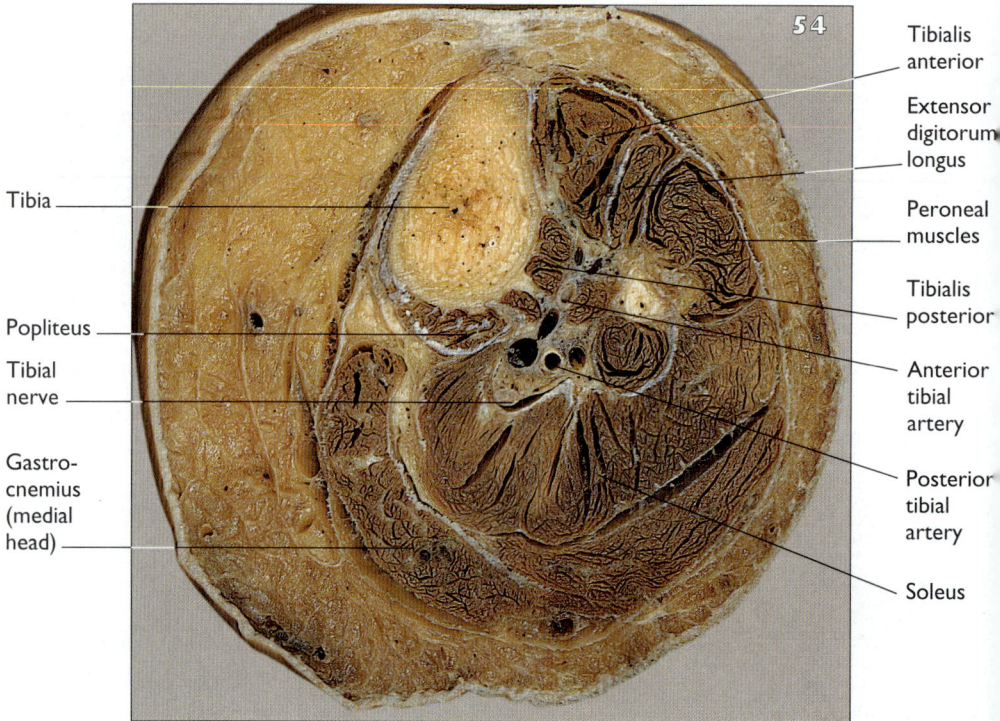

55 Dorsal view (55) of a dissection of the right foot. Identify the features labelled **A–H**.

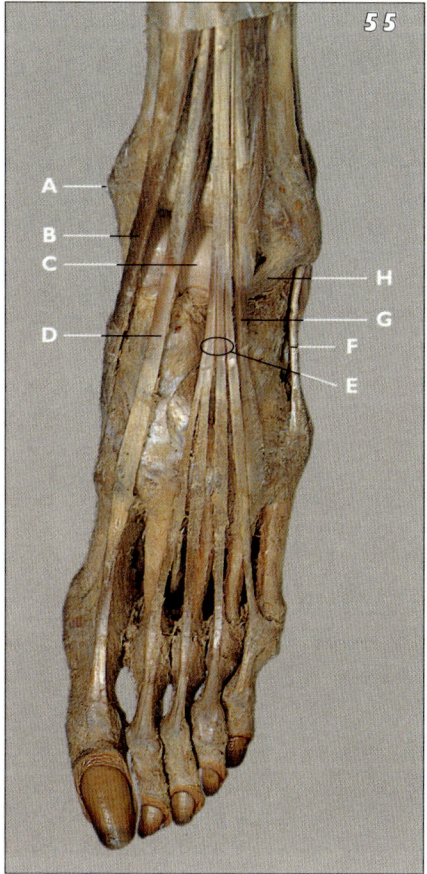

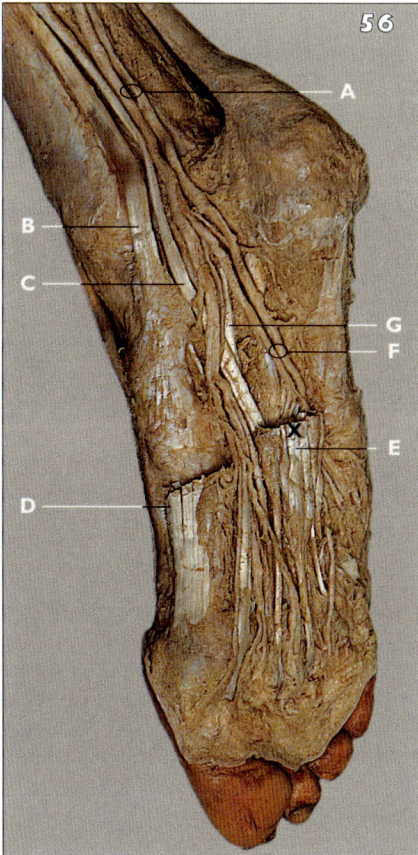

56 Dissection of the sole of the foot (56) viewed from a medial and plantar direction. Some of the muscles of the sole have been cut at **X**. Identify the structures labelled **A–G**.

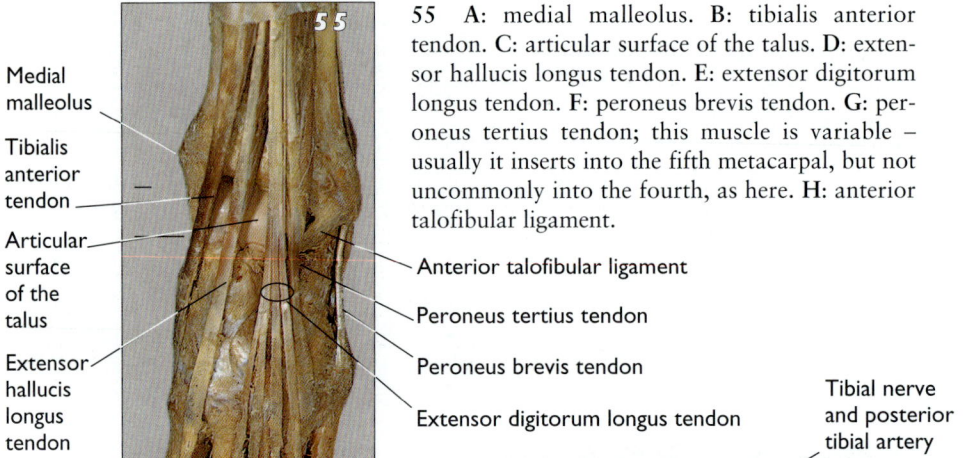

Medial malleolus

Tibialis anterior tendon

Articular surface of the talus

Extensor hallucis longus tendon

Anterior talofibular ligament

Peroneus tertius tendon

Peroneus brevis tendon

Extensor digitorum longus tendon

**55** A: medial malleolus. B: tibialis anterior tendon. C: articular surface of the talus. D: extensor hallucis longus tendon. E: extensor digitorum longus tendon. F: peroneus brevis tendon. G: peroneus tertius tendon; this muscle is variable – usually it inserts into the fifth metacarpal, but not uncommonly into the fourth, as here. H: anterior talofibular ligament.

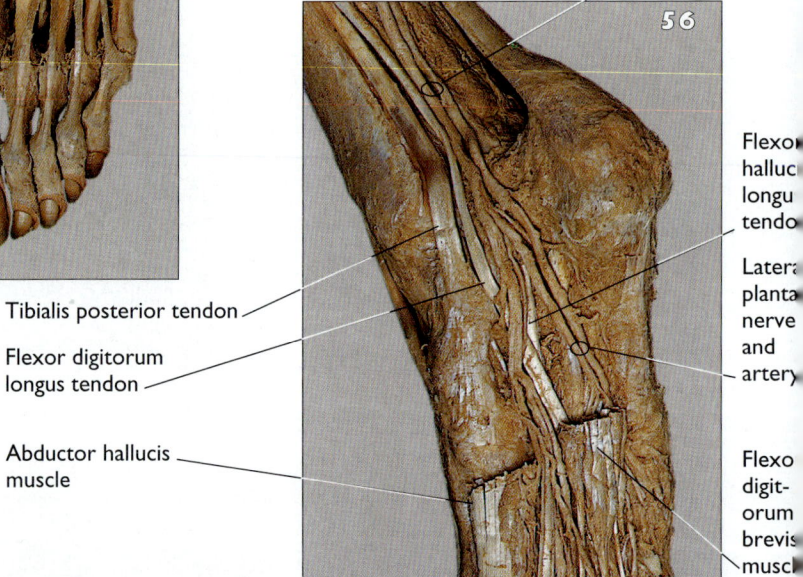

Tibial nerve and posterior tibial artery

Flexor hallucis longus tendon

Lateral plantar nerve and artery

Flexor digitorum brevis muscle

Tibialis posterior tendon

Flexor digitorum longus tendon

Abductor hallucis muscle

**56** A: tibial nerve and posterior tibial artery. B: tibialis posterior tendon. C: flexor digitorum longus tendon. D: abductor hallucis muscle. E: flexor digitorum brevis muscle. F: lateral plantar nerve and artery. G: flexor hallucis longus tendon.

57 A nurse is being trained to give an intramuscular injection into the buttock.
i. Name the large muscle that lies most superficially in the buttock.
ii. Which large nerve lies deep to this muscle and is at risk of damage from a misplaced injection?
iii. What are the muscle groups supplied by this nerve?
iv. Which areas of skin are supplied by the nerve?
v. Into which quadrant of the buttock should the nurse be taught to give the injection?
vi. Which other muscles might the needle enter if she follows this teaching?

58 A rather overweight 13-year-old boy has been complaining of mild pain on the inner side of his right knee for 2 months. Both clinical and radiological examination of the knee have been normal on two occasions. One day the boy trips and falls, and his pain becomes more severe. On examination his knee remains normal, but there is painful restriction of both active and passive movement in his left hip. Radiological examination of the hip (58) confirms the diagnosis of a slipped upper femoral epiphysis.

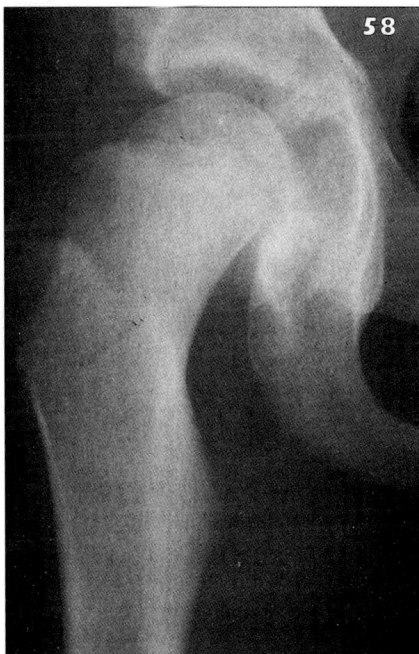

i. Why did the boy complain of knee pain rather than hip pain?
ii. What is this type of pain called?
iii. Which peripheral nerve pathway was involved in this case?
iv. Which other nerves supply the hip joint?
v. What is the general anatomical 'law' regarding the nerve supply of joints?

57 The illustrations (57a and 57b) show a dissection of the right buttock, with gluteus maximus in position and then reflected to show the underlying sciatic nerve and gluteus medius.

**i.** Gluteus maximus.

**ii.** Sciatic nerve.

**iii.** Hamstrings, flexors of ankle and toes, extensors of ankle and toes, evertors and invertors of the foot.

**iv.** The foot, except for its medial border, the back and lateral side of the leg; and a strip down the back of the thigh.

**v.** The upper and outer quadrant.

**vi.** Gluteus medius, tensor fasciae latae, or even gluteus minimus.

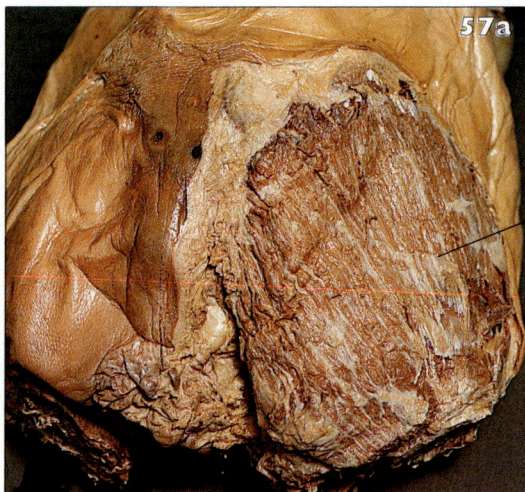

57a

Gluteus maxim▶

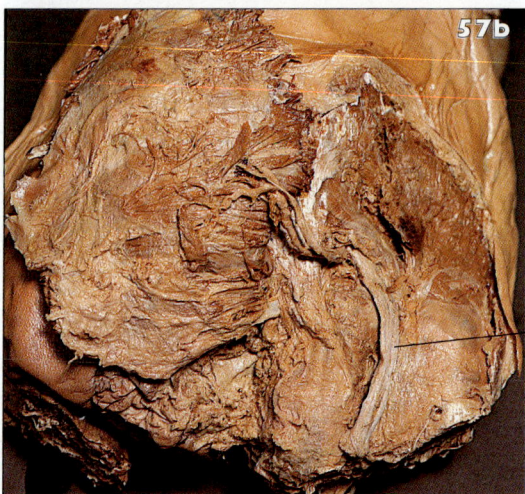

57b

Sciatic nerve

58 **i.** Because some of the nerves which supply the hip also supply the knee.

**ii.** This is an example of referred pain.

**iii.** Mainly afferent fibres in the obturator nerve.

**iv.** Sciatic and femoral nerves.

**v.** Hilton's law (John Hilton, 1805–78, Surgeon Anatomist at Guy's Hospital), which states that the nerve supplying a muscle that moves a joint, also supplies that joint and the skin over the insertion (distal attachment) of the muscle.

**59** A 28-year-old man twists his right knee while running on rough ground. The knee is initially painful, but the man is able to walk on it. The following day the knee is mildly swollen. On examination there is no definite clinical evidence of damage to any major structures within the knee. Even though the pain subsides, the man continues to have difficulty in walking. He complains that the knee feels weak and unreliable, especially on hills and stairs. On examination 2 weeks after the injury, the bulk and circumference of his right thigh are reduced compared with the left.
i. Which muscle groups are mainly responsible for the bulk of the thigh?
ii. Which of these groups plays the major role in maintaining the dynamic stability of the knee, and consequently wastes rapidly when the knee is injured or inflamed?
iii. What is the motor nerve supply of this muscle group?
iv. Which commonly examined stretch reflex involves this muscle group?
v. Which spinal nerve roots are examined when this reflex is tested?

**60** A 15-year-old girl is running downhill when her left knee suddenly collapses (gives way) and she falls to the ground. She is unable to straighten the knee fully, or to stand and walk on it. On examination, the knee is deformed, with an abnormal bony prominence laterally and tenderness and swelling anteromedially. The illustration (60) shows a magnetic resonance image of the patient's knee.
i. What is the bony abnormality? Which soft-tissue structure has been torn anteromedially?
ii. Of which type of bone is the patella the largest example?
iii. Which cutaneous nerve supplies most of the skin over the patella?
iv. Which way does the patella most commonly dislocate?
v. What is the usual anatomical reason given for the fact that patellar dislocation is more common in the female than in the male?

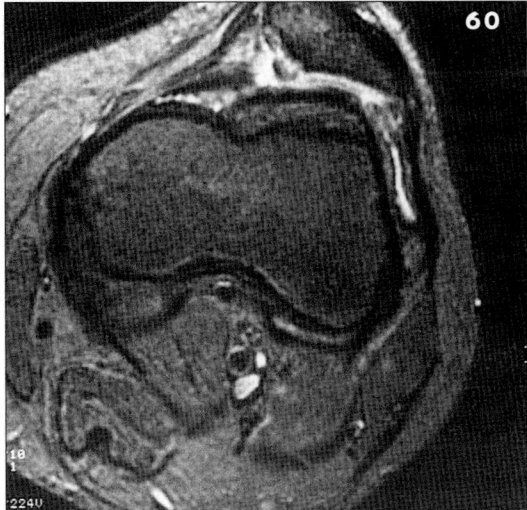

59  The illustration (59) shows an axial section through the midthigh of a right leg (seen from below), from which the relative sizes of the muscle groups can be seen.
i. Quadriceps femoris, adductors, and hamstrings.
ii. Quadriceps femoris.
iii. Femoral nerve.
iv. The knee jerk.
v. L3 and L4.

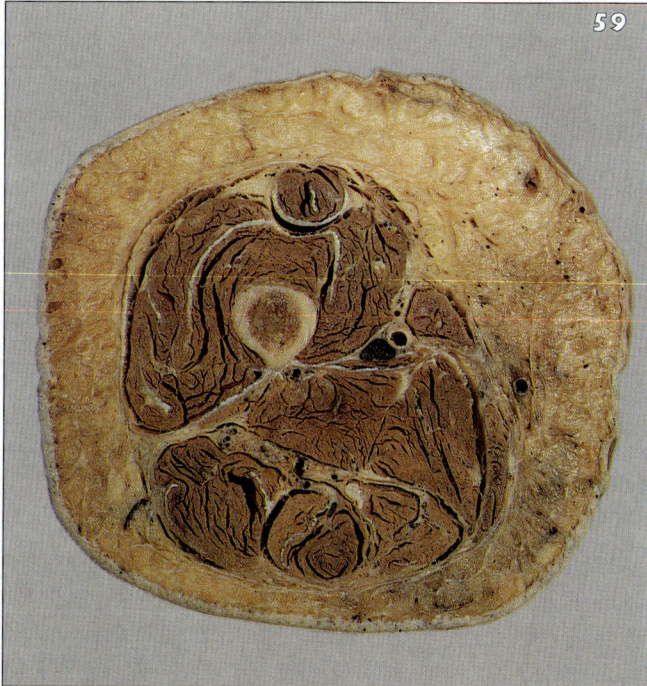

60  i. A dislocated patella. The medial patellar retinaculum.
ii. A sesamoid bone.
iii. The largest and most easily identified (and damaged) is the infrapatellar branch of the saphenous nerve, a branch of the femoral.
iv. Laterally.
v. The lateral component of the quadriceps force is larger in the female than in the male, because the angle of the femur (and hence the muscle) with the weight-bearing axis is greater in the female owing to the wider pelvis.

**61** A 75-year-old woman has had an operation to treat a fracture of the neck of her left femur. Four days after the operation, she complains of pain in the left calf. The centre of the calf is tender, and there is mild swelling around the ankle. Radiological examination (venography; **61**) confirms the clinical diagnosis of deep venous thrombosis.

**i.** Where do the main deep veins of the lower leg run?

**ii.** Which muscle of the posterior compartment of the calf contains a venous plexus?

**iii.** How do the deep veins communicate with the superficial venous system in the lower limb?

**iv.** A potentially fatal complication of deep venous thrombosis of the lower limb is embolization of a blood clot to the lungs – pulmonary embolism. Describe the anatomical pathway by which such a clot reaches the vessels of the lung.

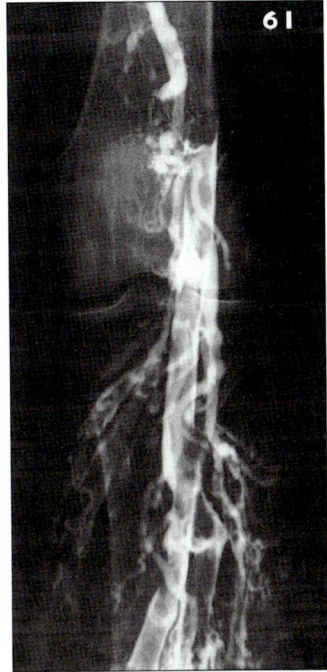

**62** A 73-year-old man, who has been a heavy smoker, complains of lack of feeling in his left big toe. On examination the whole foot is cool, and the pulp of the big toe is discoloured (**62**). There are no palpable arterial pulses in the foot. A diagnosis of ischaemia due to (peripheral) arterial disease is made.

**i.** Where, in precise anatomical terms, are arterial pulses normally felt in the lower limb during clinical examination?

**ii.** Which cutaneous nerves supply the big toe?

**iii.** Within which dermatome(s) does the big toe lie?

**iv.** Which major abdominal structure must be examined in the assessment of this patient?

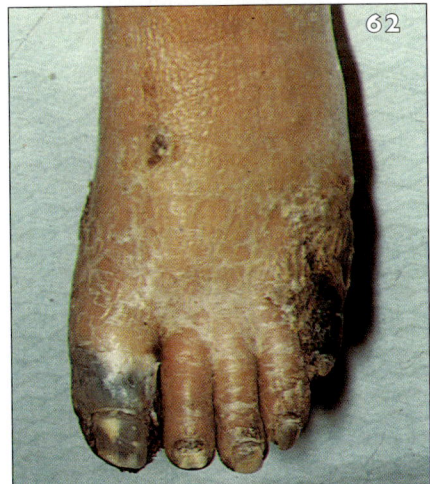

**61** **i.** With the arteries. The veins are often paired, running on either side of the artery as venae comitantes.

**ii.** Soleus. This important plexus drains mainly into the posterior tibial vein.

**iii.** The superficial great (long) saphenous vein pierces the cribriform fascia just below the inguinal ligament where it joins the deep femoral vein. The other named superficial vein the small (short) saphenous pierces the deep fascia on the back of the knee to drain into the popliteal vein. In addition, a number of perforating veins pierce the deep fascia. Important ones lie along the great saphenous vein above the medial malleolus.

**iv.** Deep veins of leg → femoral vein → external iliac → common iliac → inferior vena cava → right atrium → right ventricle → pulmonary trunk → lungs.

**62** **i.** The femoral artery, at the midpoint of the groin halfway between the ASIS and the pubic symphysis. The popliteal artery, deep in the midline of the popliteal fossa, best felt with the knee partially flexed. The dorsalis pedis is felt between the tendons of extensors hallucis and digitorum longus on the dorsum of the foot. The posterior tibial artery passes behind the medial malleolus into the medial side of the foot, almost midway between the malleolus and the Achilles tendon.

**ii.** Digital branches of the medial plantar nerve from the sole, the superficial peroneal nerve on the dorsum of the toe, and a small contribution from the deep peroneal on the lateral border of the dorsum of the toe. The saphenous nerve does not usually reach as far distally as the big toe.

**iii.** In most published diagrams it lies entirely or almost entirely within that of L4.

**iv.** The aorta, to exclude an aneurysm.

## Back

For each of questions 63–72, show whether the statement provided is true or false.

63  The pedicles of a lumbar vertebra attach to the lower half of the vertebral body.

64  All true back muscles are supplied by the ventral (anterior) primary rami of spinal nerves.

65  The posterior primary rami supply no skin.

66  The centre of a mature intervertebral disc has a blood supply from radicular arteries.

67  Interspinous ligaments are important stabilizers of the vertebral column.

68  In the midlumbar region, the erector spinae muscle group lies in an osteofascial compartment between the posterior and middle layers of the thoracolumbar fascia.

69  The abdominal wall muscles are inactive during spinal extension from vertical in the standing position.

70  Contraction of the abdominal wall muscles helps stabilize the lumbar spine via the thoracolumbar fascia.

# 63–70: Answers

63  FALSE – they attach to the upper half of the body. The lower border of the pedicle forms the superior margin of the intervertebral foramen.

64  FALSE – the true back muscles – the epaxial muscles, lying between the posterior and middle layers of the thoracolumbar fascia – are supplied by the dorsal (posterior) primary rami. The muscles lying more superficially in the back – latissimus dorsi, trapezius, the serrati – are embryologically immigrant to the back and are supplied by ventral primary rami.

65  FALSE – they supply quite a large area of the skin of the back on either side of the midline. This area widens considerably over the posterior aspects of the pectoral and pelvic girdles. Within this area, the supply remains segmental (dermatomal), as elsewhere on the trunk.

66  FALSE – in the adult, all but the periphery of the disc obtains its nutrients by diffusion from the adjacent vertebral bodies. The central area is also devoid of innervation.

67  TRUE – these ligaments are important components of the posterior column ('posterior elements') of the spine. They may be torn in flexion injuries of the spine; this is particularly important at cervical and lumbar levels. Such tears may occur in the absence of bony injury.

68  TRUE – the configuration of this compartment is of great importance to the efficient function of the muscles.

69  FALSE – the abdominal wall muscles, particularly rectus abdominis, act eccentrically to support the weight of the upper body against the force of gravity during this movement.

70  TRUE – this action is of particular importance when heavy weights are being lifted by the upper limbs.

**71** In the common form of intervertebral disc prolapse, the disc herniation lies immediately lateral to the posterior longitudinal ligament.

**72** The aorta is directly related posteriorly to the lumbar spine.

**73** Shown below is an axial section through the trunk at the level of the iliac crests (73). Identify the stuctures labelled **A–J**.

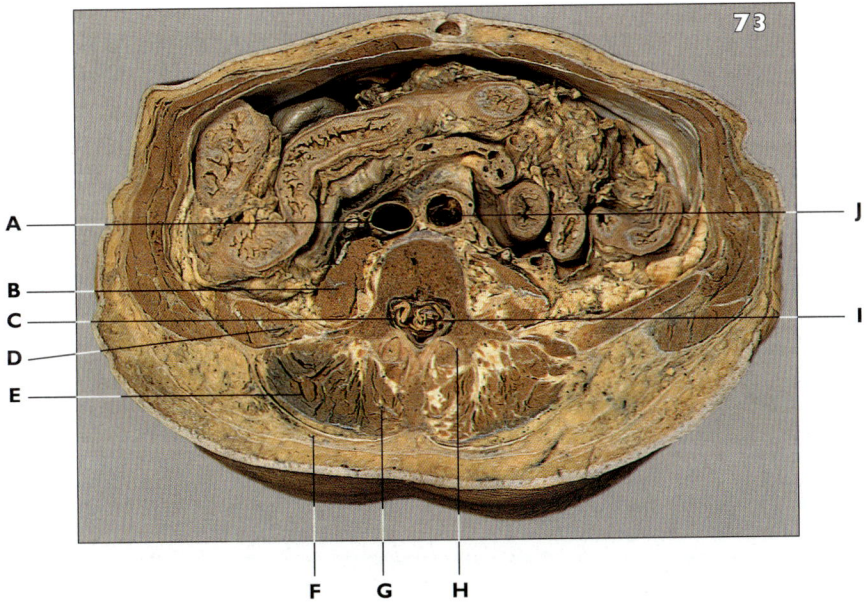

71  TRUE – the far less common, but potentially far more serious, central disc prolapse herniates or ruptures through the posterior longitudinal ligament.

72  TRUE – this is an important relationship: an aneurysm of the abdominal aorta can erode the vertebrae. The lumbar arteries leave the posterior aspect of the aorta here.

73  **A**: inferior vena cava. **B**: psoas major. **C**: dura mater (also termed the theca). **D**: quadratus lumborum. **E**: erector spinae. **F**: posterior layer of the thoracolumbar fascia. **G**: multifidus. **H**: synovial (facet or zygapophyseal) joint. **I**: cauda equina. **J**: aorta.

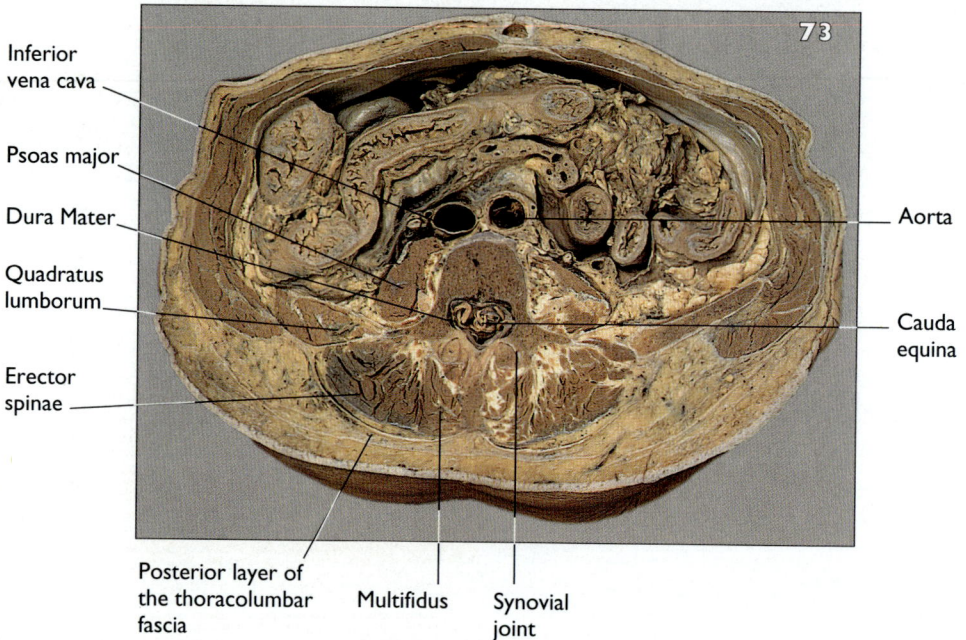

Inferior vena cava

Psoas major

Dura Mater

Quadratus lumborum

Erector spinae

Aorta

Cauda equina

Posterior layer of the thoracolumbar fascia

Multifidus

Synovial joint

74 Shown below are two photographs (74a and 74b) of a dissection of the lower back. In 74a, one layer has been folded back on the left; in 74b an additional muscle layer has been folded back. Identify the features labelled A–G.

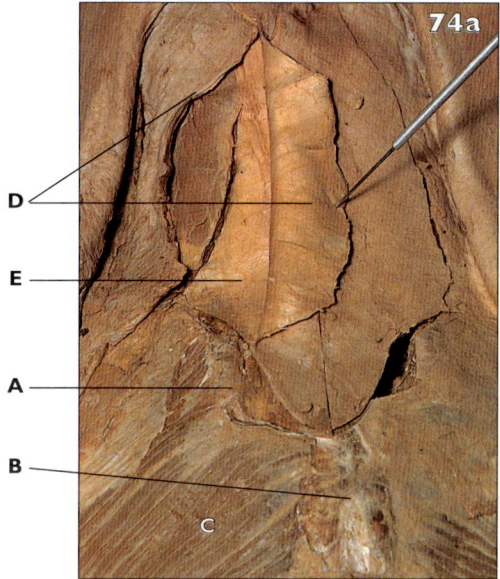

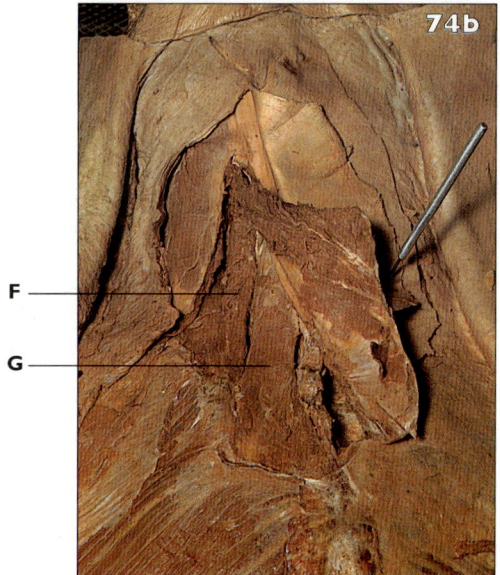

74   A: posterior superior iliac spine. B: sacral hiatus. C: gluteus maximus. D: aponeurosis of latissimus dorsi, also termed the posterior layer of the thoracolumbar fascia. E: erector spinae aponeurosis. F: erector spinae. G: multifidus.

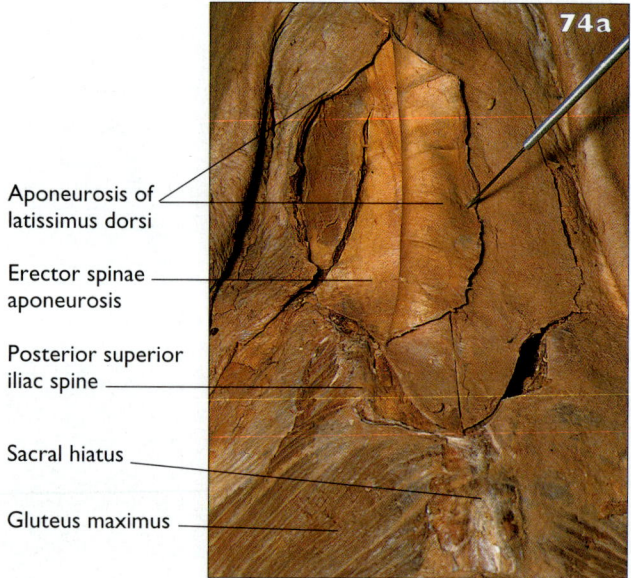

Aponeurosis of latissimus dorsi

Erector spinae aponeurosis

Posterior superior iliac spine

Sacral hiatus

Gluteus maximus

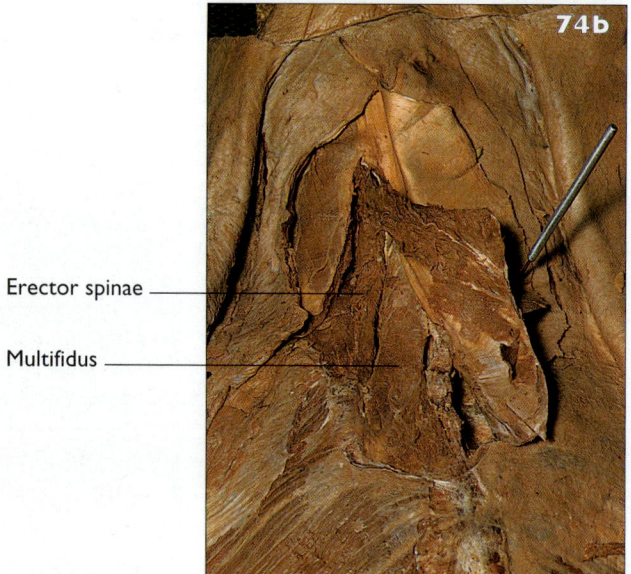

Erector spinae

Multifidus

75 A 77-year-old woman has noticed for some years that her back is becoming more bent (75a) and that her overall height is decreasing. She regularly has moderate pain in the midthoracic region of her spine after sitting or standing for long periods. There have also been episodes of more severe midthoracic pain accompanied by paraesthesiae (tingling) involving the skin of the upper abdominal wall. Radiographs of the spine have shown loss of bone density, and loss of anterior height of the vertebral bodies (wedging) together with the biconcave appearance characteristic of osteoporotic vertebral collapse.

i. Describe the curves of the normal adult vertebral column.

ii. Which of these curves is often accentuated in the osteoporotic spine?

iii. What proportion of the height of the normal adult spine is constituted by the vertebral bodies?

iv. Explain in anatomical terms the sensory symptoms involving the upper abdomen in this patient.

v. Name two other bones that commonly present with fractures in elderly osteoporotic patients.

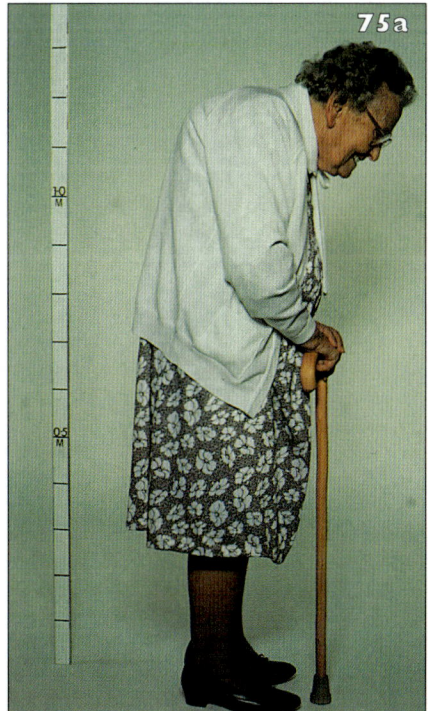

75a

**75  i.** A lordosis is a curve of the vertebral column which is concave backwards, while a kyphosis is concave forwards. The curves of the normal vertebral column are: cervical lordosis; thoracic kyphosis; lumbar lordosis; sacral kyphosis (**75b** shows the lumbosacral region of an osteoporotic vertebral column).
**ii.** In osteoporosis the bodies of the vertebrae may collapse, while the neural arches generally remain intact. This causes the vertebrae to become wedge-shaped. It most commonly affects one or more vertebrae in the thoracic region, leading to a exaggeration of the thoracic kyphosis.
**iii.** About 75% in a young adult.
**iv.** The skin of the upper abdominal wall is supplied by the midthoracic spinal nerves, particularly T7 down to T10 (which supplies the abdomen at the level of the umbilicus).
**v.** The lower end of the radius and the neck of the femur.

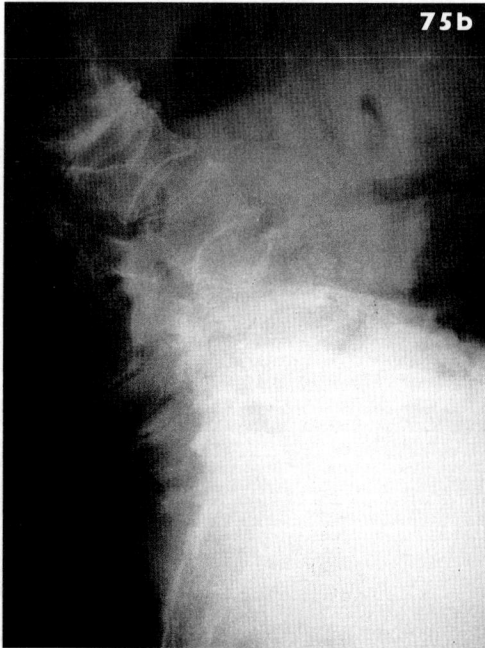

76 A 37-year-old man feels a sudden severe pain in the lower part of his back while trying to lift a pile of bricks. He has difficulty in standing straight, and some days later begins to feel pain down the back of his right lower limb and in the outer side of his right foot. On examination, forward flexion of the lumbar spine is severely limited, elevation of the straight right leg beyond 40° reproduces his pain, and there is some loss of feeling on the lateral border of the right foot. A clinical diagnosis of prolapsed intervertebral disc with nerve root irritation is made and confirmed radiologically by a CT scan (76).

i. Which nerve root is involved?

ii. Where does this root leave (exit from) the spine?

iii. How would you examine the myotome (motor area of supply) of this root?

iv. Which disc is likely to have prolapsed?

v. How is the usual, 'lateral', disc prolapse related anatomically to the posterior longitudinal ligament?

vi. What is the particular anatomical danger of a 'central' disc prolapse?

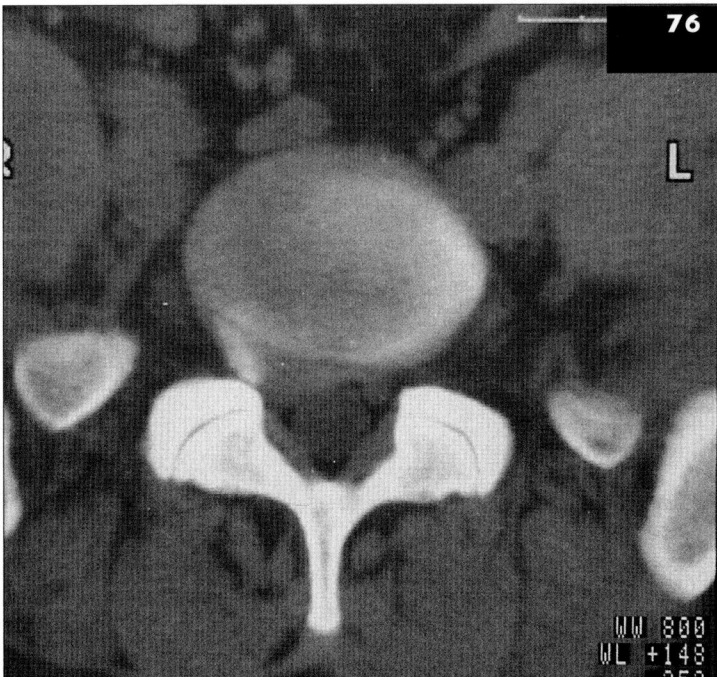

76  i. The first sacral (S1).

ii. Through the first foramen on the ventral surface of the sacrum.

iii. Test the ankle jerk and the power of plantar flexion of the ankle.

iv. The lumbosacral disc (between L5 and S1).

v. The disc herniates immediately lateral to the ligament.

vi. Damage to the sacral (S2–S4) nerve roots, resulting in loss of supply to the pelvic viscera as well as paralysis of lower leg and foot muscles.

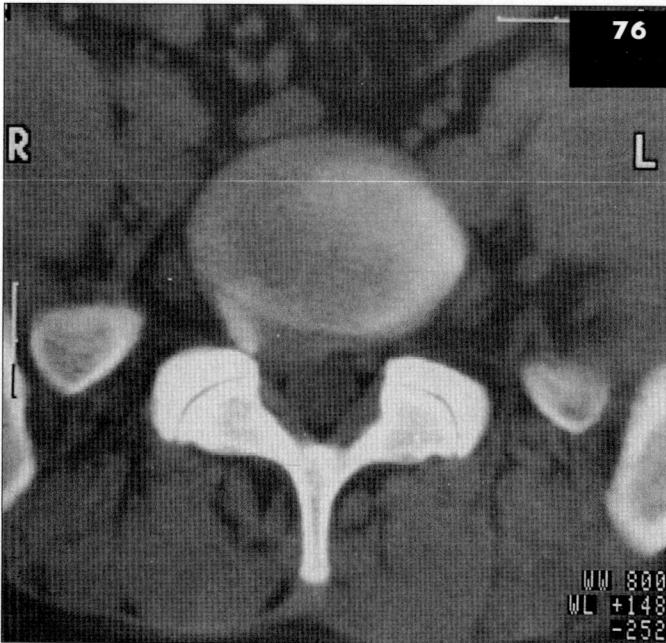

# THE CARDIOVASCULAR AND RESPIRATORY SYSTEMS

## Thorax

For each of questions 77–96, show whether the statement provided is true or false.

77  A typical rib articulates with vertebrae by secondary cartilaginous joints.

78  The joint between the manubrium and the body of the sternum is usually a primary cartilaginous joint.

79  The eleventh rib has no costal cartilage.

80  The manubrium sterni shows articular facets for the first three costal cartilages.

81  The internal intercostal muscles are replaced anteriorly by aponeurotic membranes.

82  The posterior intercostal arteries all branch directly from the descending aorta.

83  The posterior intercostal veins all drain directly into the azygos vein.

84  The healthy lung is highly elastic, pink, and floats in water.

85  The parietal and visceral layers of the pleura become continuous around the hilum of the lung.

77  FALSE – a typical rib articulates with thoracic vertebrae by means of synovial joints. For example, the head of the fifth rib articulates with demifacets on the fourth and fifth thoracic vertebral bodies. In addition, the tubercle on the neck of the fifth rib articulates with a facet on the transverse process of the fifth thoracic vertebra.

78  FALSE – the manubriosternal joint, being a midline joint, is usually a symphysis (i.e. a secondary cartilaginous joint, with a central fibrous component). With age, the joint may ossify.

79  FALSE – all the ribs have costal cartilages and develop endochondrally. The costal cartilages of ribs 11 and 12 – the 'floating' ribs – do not articulate with the sternum or the cartilages of the ribs above.

80  FALSE – only the first and second costal cartilages articulate with the manubrium sterni (the first by means of a primary cartilaginous joint, the second via a synovial joint). The clavicle also articulates with the manubrium sterni by means of a synovial joint.

81  FALSE – the external intercostal muscles are replaced anteriorly, near the sternum, by aponeurotic membranes; the internal intercostal muscles are replaced posteriorly, near the spine, by membranes.

82  FALSE – the first and second arise from the superior intercostal artery (costo-cervical trunk).

83  FALSE – most of the posterior intercostal veins on the right side drain into the azygos vein; posterior intercostal veins on the left drain only indirectly via hemiazygos and/or accessory hemiazygos veins. The posterior intercostal veins in the first intercostal spaces can drain into the brachiocephalic veins.

84  TRUE – most city-dwellers will have black carbon deposits within their lungs. The elastic characteristics are important during expiration. The property of floating in water might have forensic value: for example, the lungs of a stillborn baby will sink.

85  TRUE – the visceral pleura covers the surface of the lung; the parietal pleura covers the walls of the pleural cavity. The lungs 'invaginate' the pleura so that the hila connect with the mediastinum and are bounded by a 'cuff' of pleura consisting of both layers.

86 Pleural fluid decreases friction during respiratory movements.

87 The intrapleural pressure is greater than atmospheric throughout respiration.

88 The left lung has three lobes.

89 The apices of the lungs extend into the root of the neck.

90 The 'plane of Louis' extends from the lower border of the manubrium sterni to the lower border of the sixth thoracic vertebra.

91 The ligamentum arteriosum lies at the level of the 'plane of Louis'.

92 The apex beat of the normal heart is best felt in the left fifth intercostal space, just medial to the midclavicular line.

93 The posterior surface (base) of the heart consists mainly of the right atrium.

94 Both the right and left coronary arteries arise just above the margins of pulmonary semilunar valves.

# 86–94: Answers

86 TRUE – note that, clinically, scarring of the pleura, after injury or infection, may lead to adhesions between the lung and the chest wall.

87 FALSE – the intrapleural pressure is often said to be a 'negative pressure'. There is, of course, no such thing as a negative pressure and the implication is that the pressure is subatmospheric. Because of this, the lungs, being at atmospheric pressure, are held distended. If the pleura is punctured in such a way that air enters the pleural cavity and the intrapleural pressure becomes atmospheric, the lungs will collapse; this is termed pneumothorax.

88 FALSE – the right lung has three lobes (demarcated by oblique and transverse fissures); the left lung, lacking a transverse fissure, has only two lobes.

89 TRUE – trauma at the root of the neck can cause pneumothorax (see answer 87). A carcinoma within the apex of the lung might impinge upon structures in the root of the neck, such as the first thoracic spinal nerve (which, because of its sympathetic fibres, might produce derangement of the autonomic system in the head and neck – Horner's syndrome [see answers 162 and 182]) and the brachial plexus (causing problems in the upper limb).

90 FALSE – the 'plane of Louis' extends from the manubriosternal joint to the fourth thoracic vertebra. It is a plane that demarcates the lower border of the superior mediastinum and is a landmark for many structures within the mediastinum.

91 TRUE – the ligamentum arteriosum is the remains of the foetal ductus arteriosus that shunts blood from the pulmonary circulation to the systemic circulation (the aorta) before birth. It is also the site where the left recurrent laryngeal nerve branches from the left vagus and hooks behind the arch of the aorta to ascend to the larynx.

92 TRUE – this is also the best site for hearing the mitral valve.

93 FALSE – the right atrium is located more along the right border of the heart; the left atrium occupies most of the posterior surface (base) of the heart.

94 FALSE – the right coronary artery arises above the anterior cusp of the aortic semilunar valve; the left coronary artery arises in association with the left posterior cusp of the aortic semilunar valve.

95  The left coronary artery lies within the anterior atrioventricular groove.

96  The coronary sinus, draining venous blood from the heart, opens into the left atrium.

97  Identify the labelled structures (A–J) on the photograph (97) of the left side of the thoracic cavity.

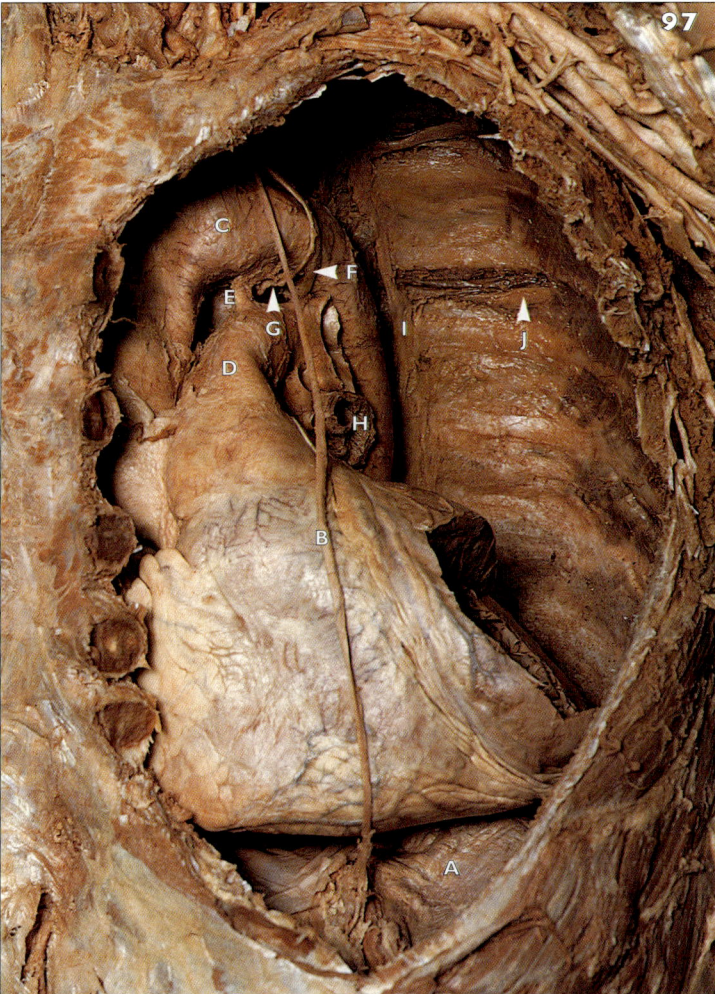

95 FALSE – the right coronary artery runs within the anterior atrioventricular groove.

96 FALSE – the coronary sinus drains into the right atrium of the heart.

97 **A:** diaphragm. **B:** left phrenic nerve. **C:** arch of aorta. **D:** pulmonary trunk. **E:** ligamentum arteriosum. **F:** left vagus nerve (cranial nerve). **G:** left recurrent laryngeal nerve. **H:** left main bronchus. **I:** left thoracic sympathetic trunk. **J:** intercostal nerve.

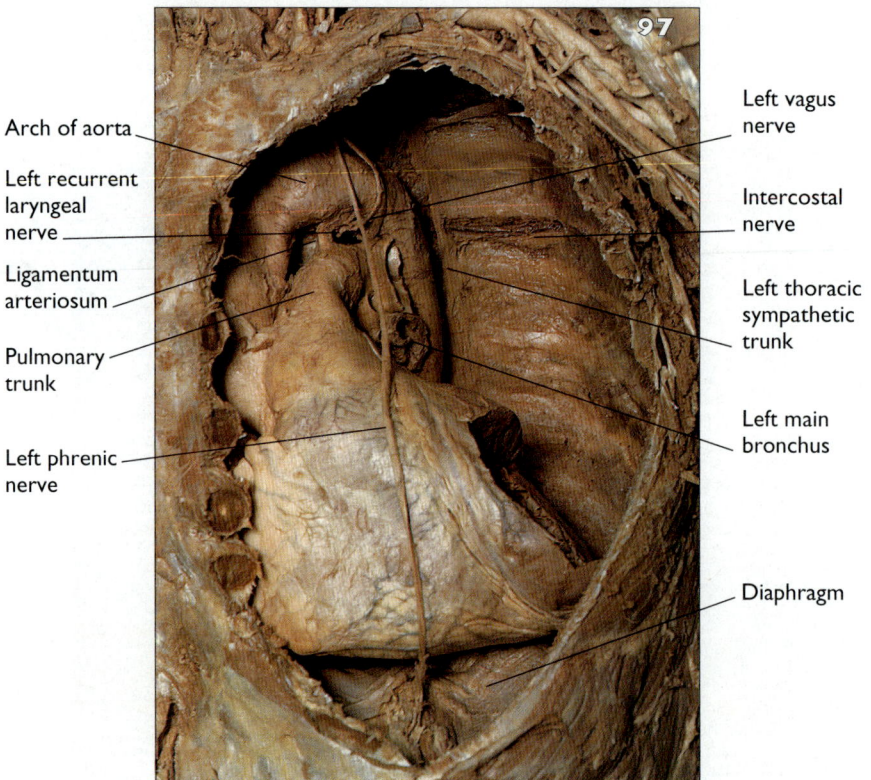

98  Identify the labelled structures **A–H** on the photograph of the anterior surface of the heart (**98**).

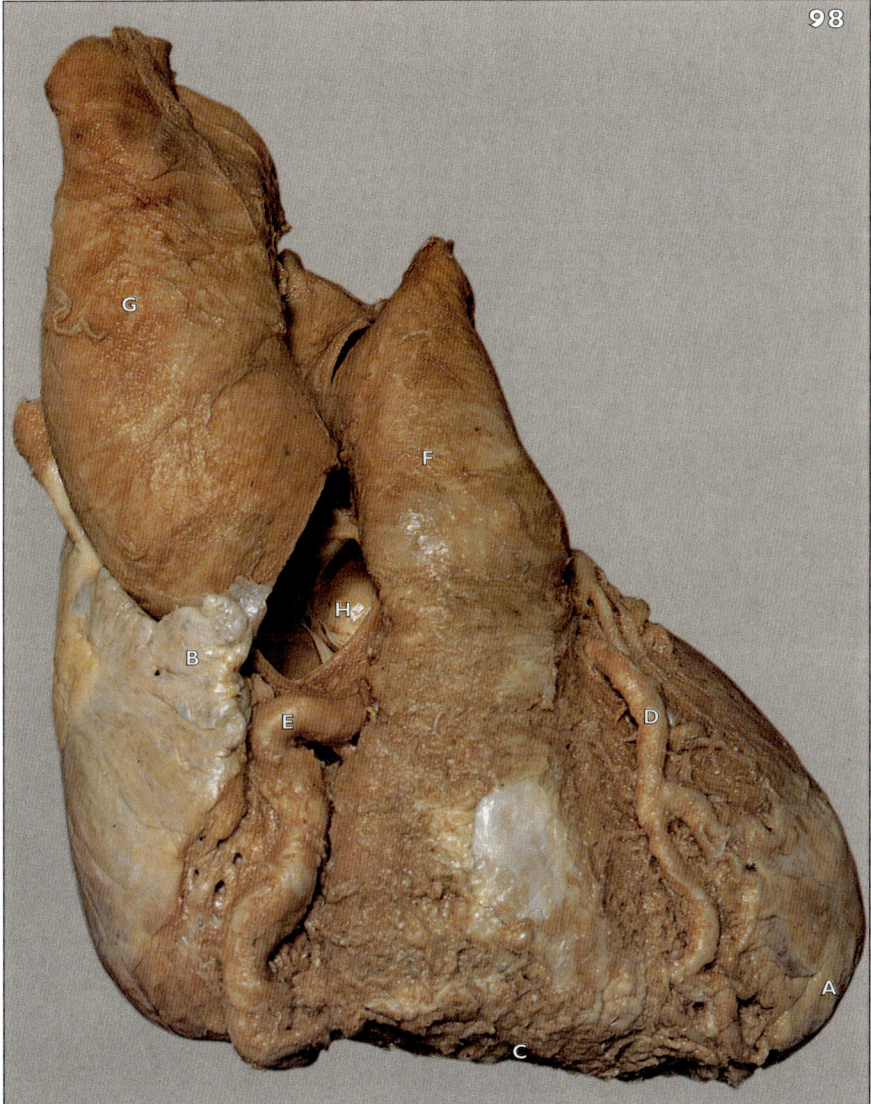

98   **A:** apex of heart. **B:** right auricle. **C:** diaphragmatic surface of heart. **D:** anterior interventricular branch of left coronary artery. **E:** right coronary artery. **F:** pulmonary trunk. **G:** ascending aorta. **H:** semilunar aortic valve.

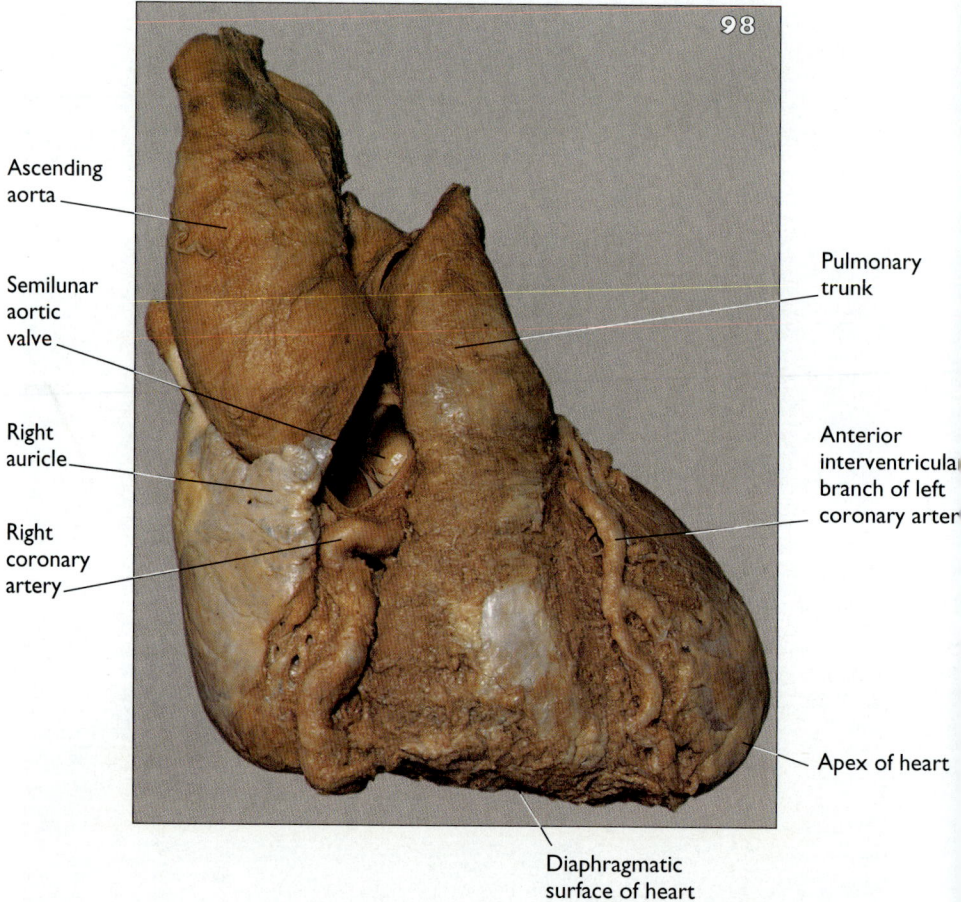

Ascending aorta

Semilunar aortic valve

Right auricle

Right coronary artery

Pulmonary trunk

Anterior interventricular branch of left coronary artery

Apex of heart

Diaphragmatic surface of heart

99 Identify the labelled structures **A–J** on the photograph of the posterior surface of the heart (**99**).

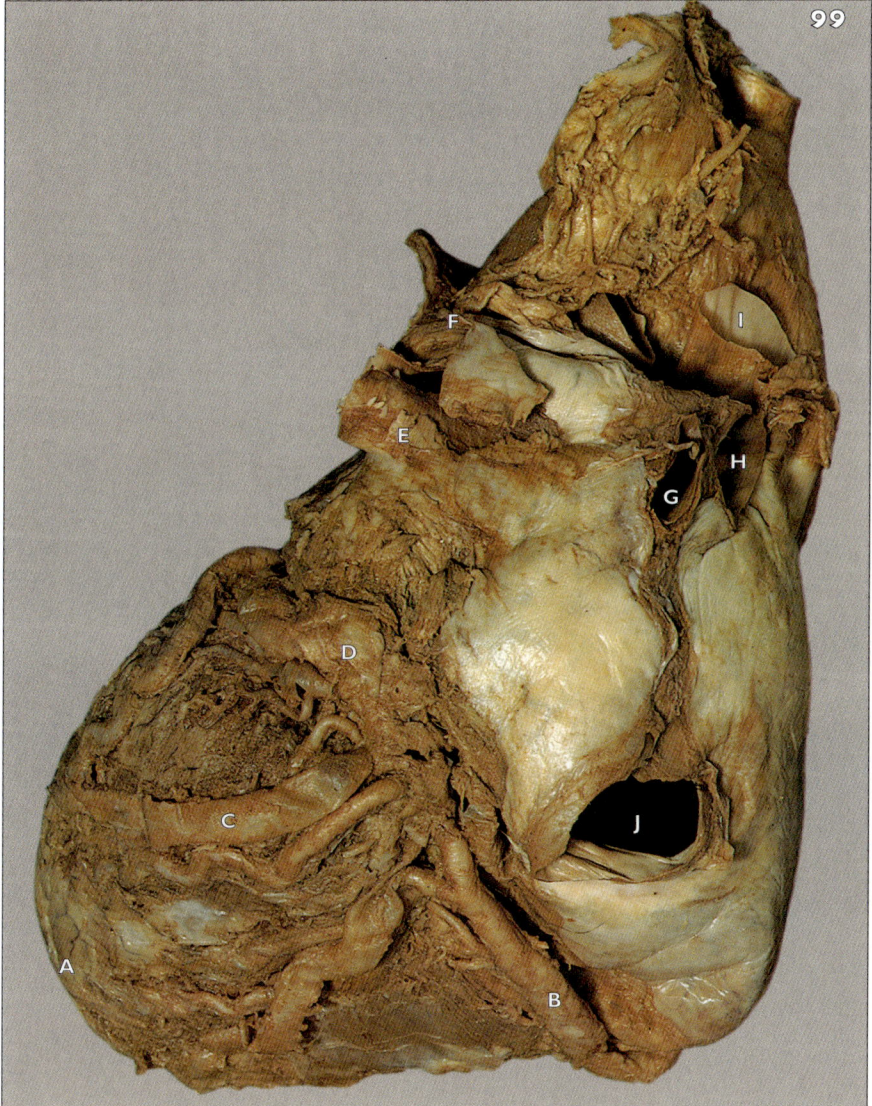

# 99: Answers

**99** **A:** apex of heart. **B:** right coronary artery. **C:** great cardiac vein. **D:** coronary sinus. **E:** left inferior pulmonary vein. **F:** left superior pulmonary vein. **G:** right inferior pulmonary vein. **H:** right superior pulmonary vein. **I:** superior vena cava. **J:** inferior vena cava.

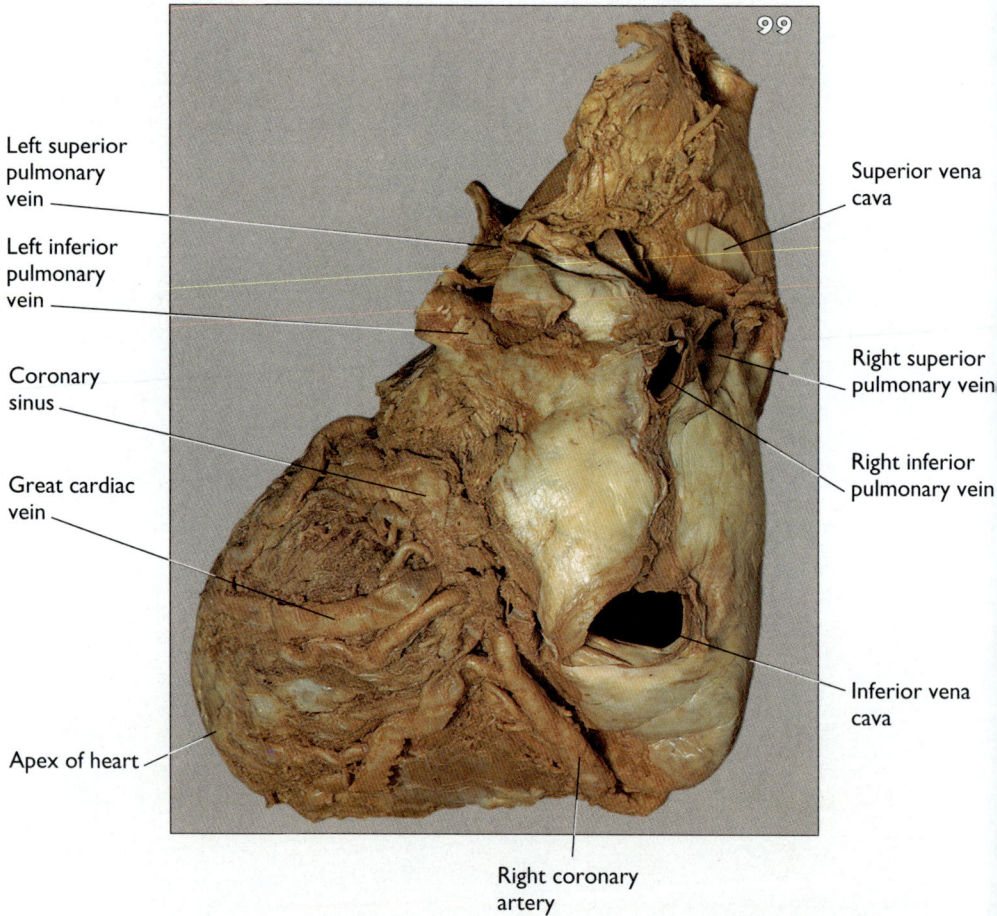

Left superior pulmonary vein

Left inferior pulmonary vein

Coronary sinus

Great cardiac vein

Apex of heart

Superior vena cava

Right superior pulmonary vein

Right inferior pulmonary vein

Inferior vena cava

Right coronary artery

**100** Identify the labelled structures **A–J** on the photograph of an axial (transverse) section through the thorax (**100**).

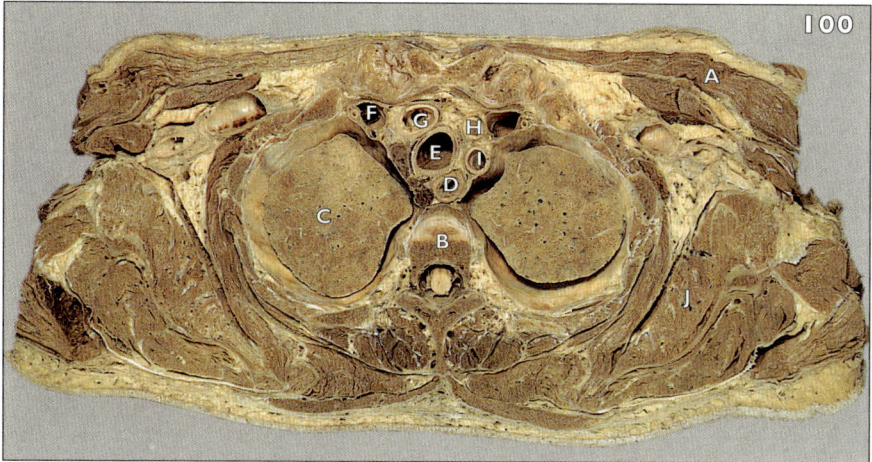

**101** Identify the labelled structures **A–L** on the photograph of an axial (transverse) section through the thorax (**101**).

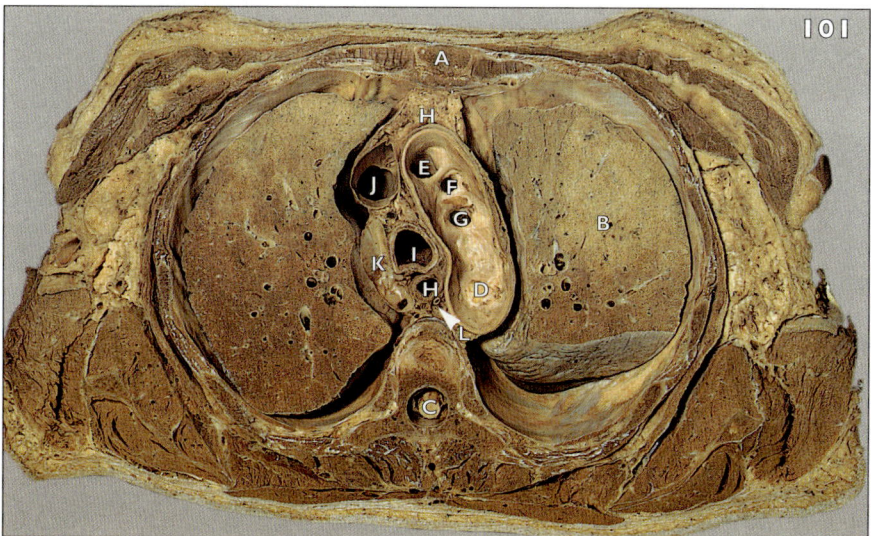

**100**  A: pectoralis major. B: body of second thoracic vertebra. C: apex of right lung.
D: oesophagus. E: trachea. F: right brachiocephalic vein. G: brachiocephalic artery. H:
left common carotid artery. I: left subclavian artery. J: subscapularis.

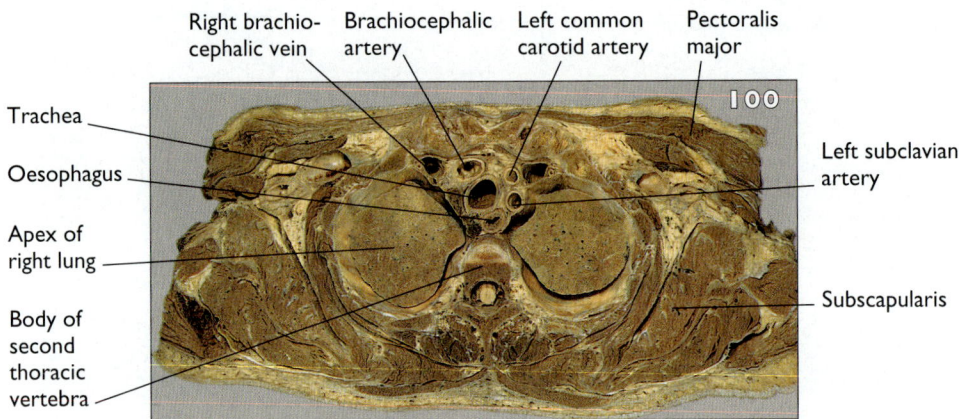

**101**  A: body of sternum. B: left lung. C: thoracic spinal cord. D: arch of aorta.
E: brachiocephalic artery. F: left common carotid artery. G: left subclavian artery.
H: oesophagus. I: trachea. J: superior vena cava. K: azygous vein. L: thoracic duct.

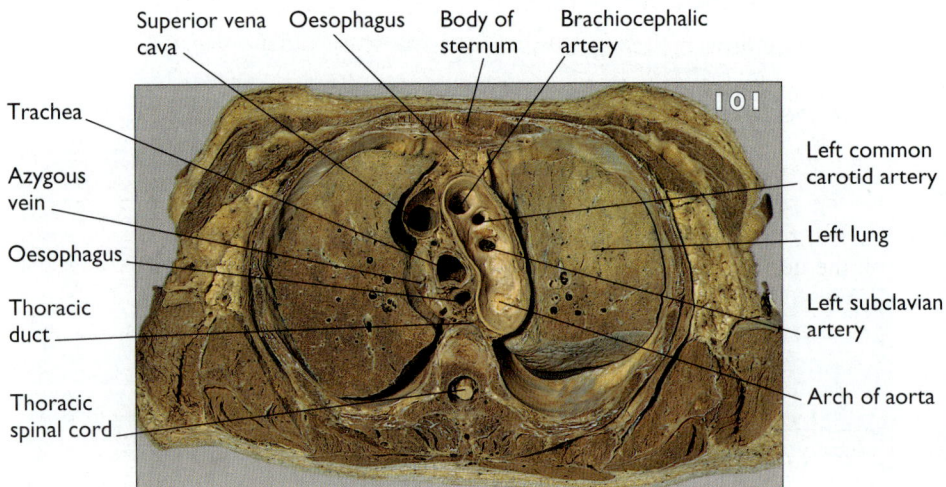

**102** A 20-year-old rugby player is heavily tackled and subsequently develops pain on the right side of his chest. In the trauma department of the local hospital, a chest radiograph (**102**) shows fracturing of the right third rib. An intercostal nerve block will be effective in alleviating the pain. Apart from 'blocking' the intercostal nerve of the fractured rib, some intercostal nerves above and below the fractured rib should also be anaesthetized.

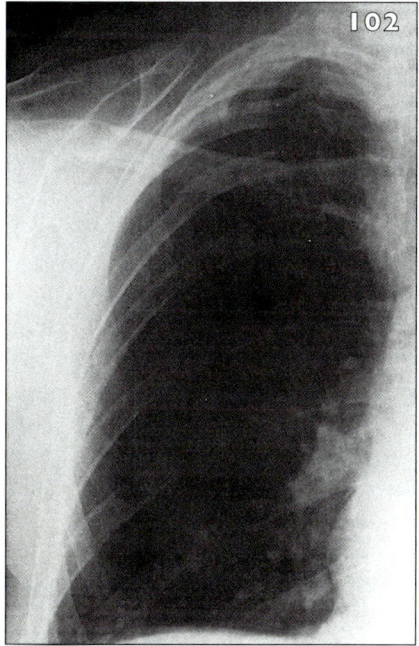

**i.** For the third intercostal nerve, how is the needle of the anaesthetic syringe directed and what structures would the needle cross in order to anaesthetize the intercostal nerve?

**ii.** If the injection is poorly accomplished, what underlying thoracic structures might be damaged?

**103** A 20-year-old woman suffers from Raynaud's disease. This consists of spasm of the arteries supplying the fingers, and is precipitated by cold and relieved by heat. For this patient it causes numbness and a burning sensation in the fingers, with severe pain in the rewarming phase. Sympathetic denervation (sympathectomy) may be necessary in this case and it is decided by the surgeons to adopt a 'transaxillary approach', the skin being incised over the third rib on the thoracic wall of the axilla (**103a**). Such an approach, with reflection of the lung, will gain access to the second and third (possibly fourth) thoracic sympathetic ganglia on the necks of the ribs.

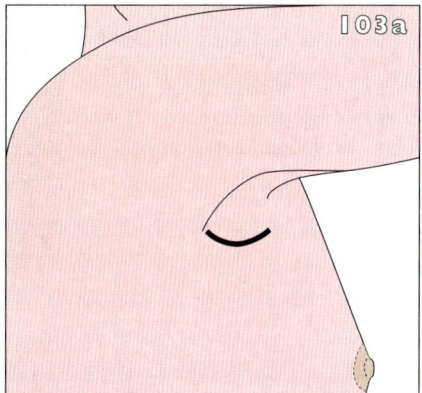

**i.** Which muscles must be reflected or pierced to enter the thorax?

**ii.** Which nerves and vessels running in the thoracic wall of the axilla should be avoided with the incisions (what would be the effect of damaging the nerves)?

**102  i.** The third intercostal nerve lies within the costal groove of the third rib. It therefore lies under the lower border of the rib. Consequently, the needle of the anaesthetic syringe has to be directed upwards and under the third rib. In the process of getting the needle to the nerve, the skin and the external and internal intercostal muscles have to be traversed.

**ii.** If the needle passes beyond the nerve and penetrates the innermost intercostal muscle layer, it could penetrate the pleura and the lung; this could produce a pneumothorax (see **87**).

**103  i.** The muscles to be reflected or pierced are pectoralis major, latissimus dorsi, and serratus anterior (**103b**).

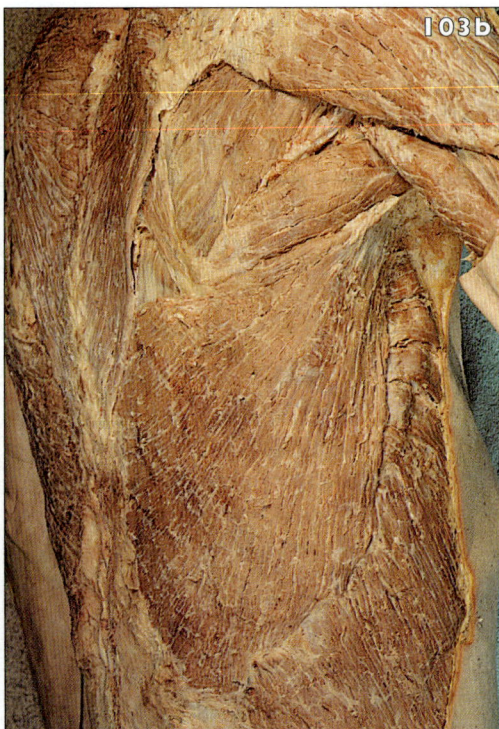

**ii.** The long thoracic nerve, the thoracodorsal nerve, and the subscapular vessels are at risk from a transaxillary incision. Damage to the long thoracic nerve will result in the loss of motor supply to serratus anterior; as a consequence, the trunk will fall forwards on that side when the weight is taken on the arms and the scapula will appear to be 'winged'. Damage to the thoracodorsal nerve will result in loss of motor supply to latissimus dorsi; as a consequence, there will be loss of adduction, extension and medial rotation of the upper arm.

**104** A 50-year-old smoker presents to his physician with weight loss and a persistent cough. He is referred to a cardiothoracic surgeon following the discovery of a large mass in the lung field on a chest radiograph (**104a**). There is no evidence of metastasis and the surgeon recommends removal of the left lung in the hope of improving the patient's survival. The procedure for the removal of a lung is known to surgeons as a pneumonectomy. Essentially, it requires the severing of all structures at the hilum.

**i.** Describe the left and right hila of the lungs, emphasizing differences.

**ii.** For a left pneumonectomy:

(a) The left bronchus is difficult to close because of which adjacent blood vessel?

(b) Which two nerves running on either side of the left hilum could be damaged during surgery?

**iii.** For a right pneumonectomy:

(a) Which blood vessel arching over the right hilum must be protected (assuming that it is not pathologically involved)?

(b) Sectioning structures in the right hilum from superior to inferior, list the order of structures to be dissected out and divided.

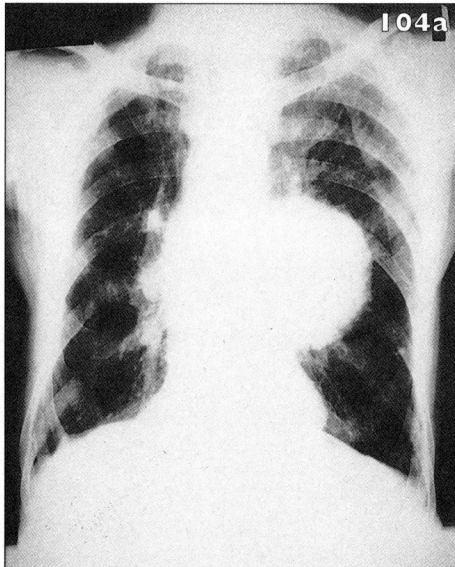

**104  i.** The hilum (root) of the lung carries the major blood vessels and bronchi from the mediastinum and into the lung. It is surrounded by a 'cuff' of pleura. The general arrangement of structures within a hilum, from superior to inferior, is: pulmonary artery and bronchus, pulmonary veins. On the right, the bronchus divides into upper lobe bronchus and right main bronchus before it enters the lung. The upper lobe bronchus is also called the eparterial bronchus because it passes above the main pulmonary artery. The pulmonary artery on the right also divides before entering the lung. The illustration (**104b**) highlights the differences between the right and left hila. Note that bronchial arteries also pass through the hilum and that there is a pulmonary nervous plexus surrounding the bronchi. Lymph nodes may also be found.

**ii. (a)** The left bronchus is difficult to close because of the close proximity of the arch of the aorta.

**(b)** Passing in front of the left hilum, the left phrenic nerve, passing behind the hilum, the vagus nerve.

**iii. (a)** The azygos vein.

**(b)** The structures to be sectioned in the right hilum are, as stated above, the pulmonary arteries, bronchi, and the pulmonary veins.

RIGHT        LEFT

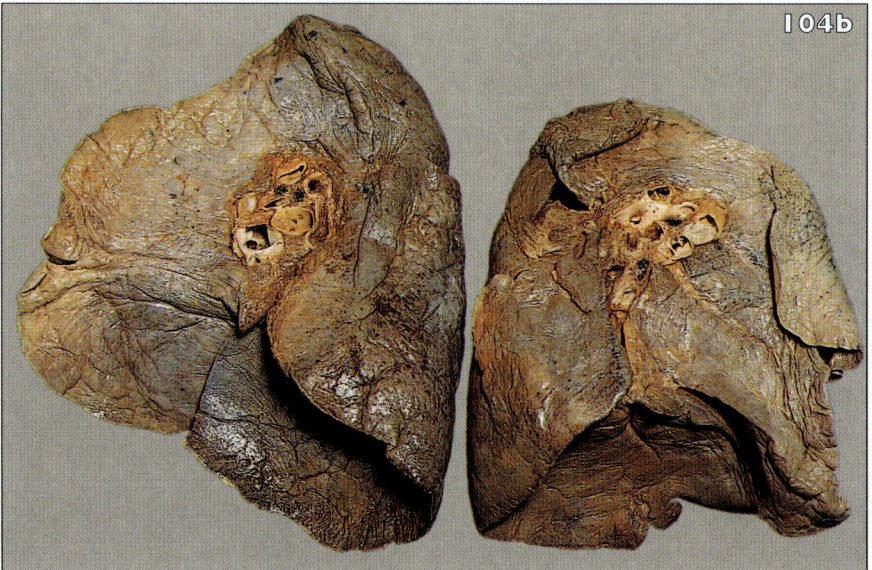

104b

105 A 60-year-old patient, who admitted to being a very heavy smoker, went to his physician complaining of a worsening cough, hoarseness of voice, and blood in his sputum. The physician noted that there was a firm swelling on the right side of the root of the neck near the midline. A chest radiograph (105a) revealed irregular, radiopaque masses within the right and left lungs.

i. Why do patients with primary carcinoma of the left lung near the hilum often present with a hoarse voice and why is hoarseness unlikely with a primary carcinoma of the right lung?

ii. On the basis of the case history and the radiological evidence, the patient was diagnosed as having secondary malignant lesions within the lungs, stemming from a primary carcinoma in the thyroid gland. How would thyroid carcinoma cause breathlessness (dyspnoea)?

iii. The tumour may spread from the thyroid gland via the lymphatic system. If so, where will it spread?

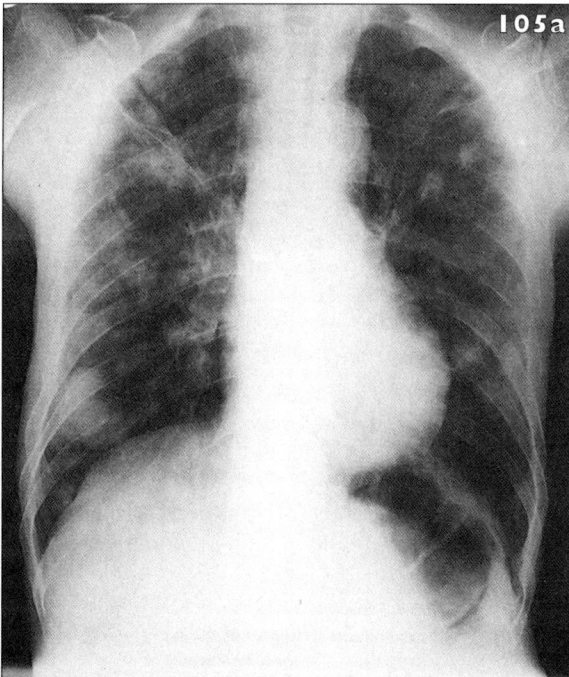

# 105: Answers

**105** **i.** On the left side, the left recurrent laryngeal nerve (a nerve that passes upwards from the thorax to the larynx in the neck to supply the left vocal cord) originates from the left vagus nerve in the mediastinum and runs close to the left main bronchus. On the right side, however, the recurrent laryngeal nerve does not appear in the thorax but arises from the right vagus near the subclavian artery in the root of the neck. The reason for the asymmetry relates to the development of the great vessels in the embryo (**105b**). The recurrent laryngeal nerves are associated with the sixth branchial arches: on the left, the sixth aortic arch develops into the ductus arteriosus and therefore the recurrent laryngeal nerve associated with this remains in the thorax; on the right, however, the distal parts of the fifth and sixth aortic arches are absent and, consequently, the right recurrent laryngeal nerve becomes associated with the fourth aortic arch, which goes on to develop into the right subclavian artery.

**ii.** Local spread takes place in thyroid carcinoma, with compression and invasion of the trachea. The proximity of the recurrent laryngeal nerves to the thyroid gland can also produce hoarseness.

**iii.** Numerous lymphatic vessels leave the gland and drain to lymph nodes situated: (**a**) in the midline on the front surface of the larynx and trachea, (**b**) in the tracheo-oesophageal groove extending downwards into the superior mediastinum, and (**c**) upwards and laterally to the deep cervical chain and to nodes lateral to this.

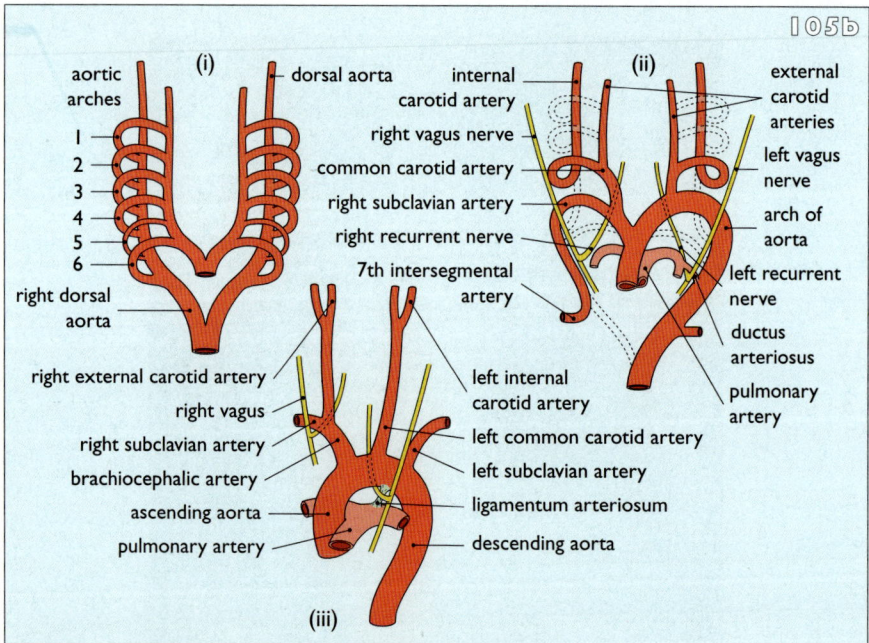

**i. Aortic arches and dorsal aortas before 'transformation'.**
**ii. Aortic arches and dorsal aortas after 'transformation'.**
**iii. The great arteries and vagus and recurrent laryngeal nerves in the adult.**

106 A 45-year-old woman complains of heartburn, especially after meals. This is made worse by stooping and lying down. She is investigated radiologically with a barium meal (106a), which demonstrates a hiatus hernia. In this condition, part of the stomach herniates through the oesophageal hiatus of the diaphragm.

i. Through which part of the diaphragm does the oesophagus enter the abdomen, and at what level does this occur?

ii. Which other structures are transmitted through the oesophageal opening of the diaphragm?

iii. What features prevent reflux of stomach contents back into the oesophagus?

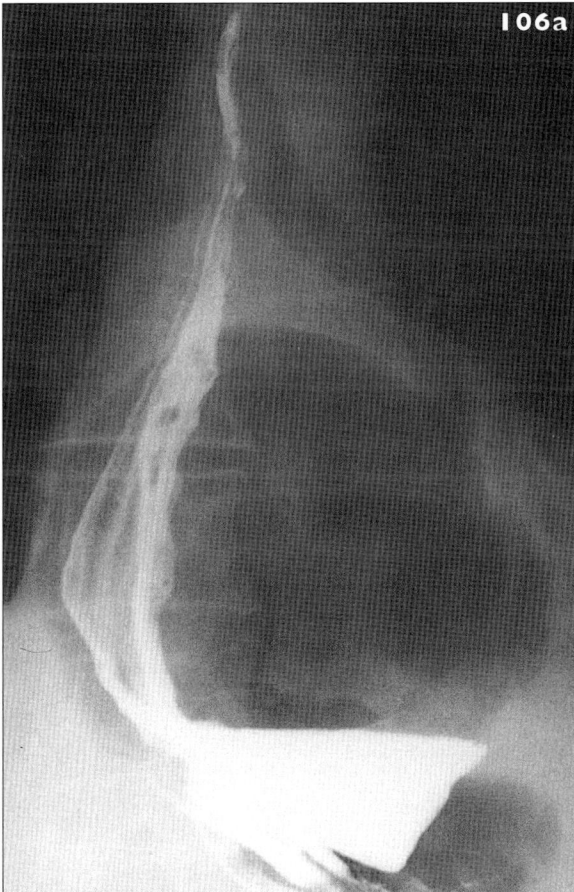

**106 i.** The oesophageal opening is in the right crus of the diaphragm (**106b**), but it is slightly to the left of the median plane. The opening is at the level of the tenth thoracic vertebral body.

**ii.** Passing through the oesophageal opening of the diaphragm, in addition to the oesophagus, are the right and left vagi and the oesophageal branches of the left gastric artery with accompanying veins.

**iii.** Reflux of stomach contents into the oesophagus has deleterious effects upon the oesophageal epithelium. There is some controversy as to how reflux is prevented. A cardiac sphincter has been described as existing at the point where the oesophagus enters the stomach, but for some anatomists this sphincter is non-existent. A functional sphincter might be produced by combination of the right crus of the diaphragm, the circular muscle fibres in the oesophagus, the angle between the oesophagus and the fundus of the stomach, and the folds of mucosa.

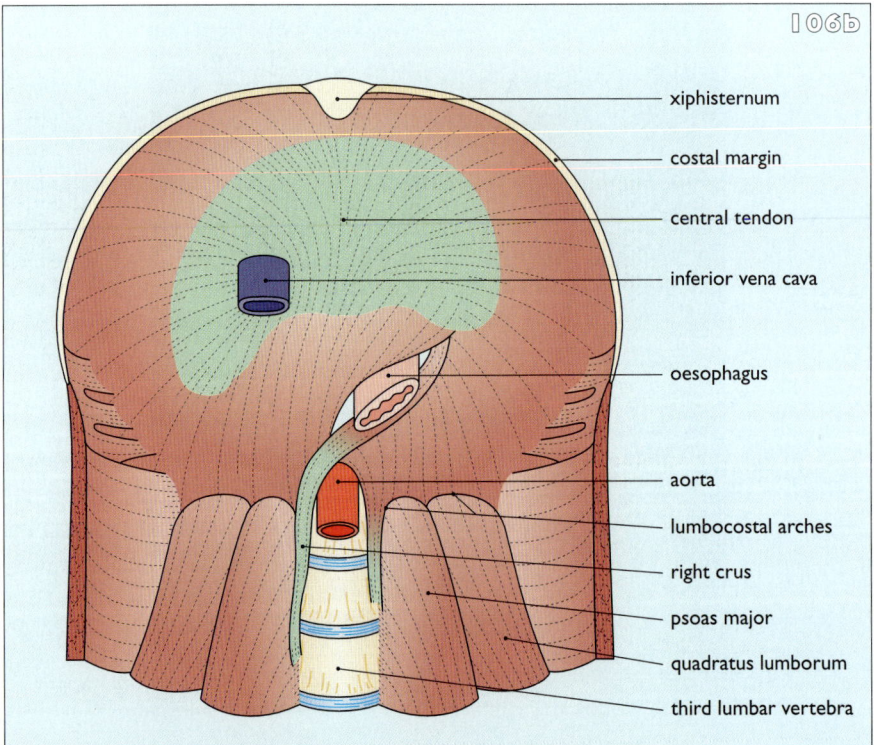

106b

- xiphisternum
- costal margin
- central tendon
- inferior vena cava
- oesophagus
- aorta
- lumbocostal arches
- right crus
- psoas major
- quadratus lumborum
- third lumbar vertebra

**Abdominal surface of the diaphragm**

107  The illustration below (107) is a precardial echocardiogram showing a long-axis parasternal cross-section of a normal heart. Identify the structures labelled 1–8. During which part of the cardiac cycle was this image obtained, and why?

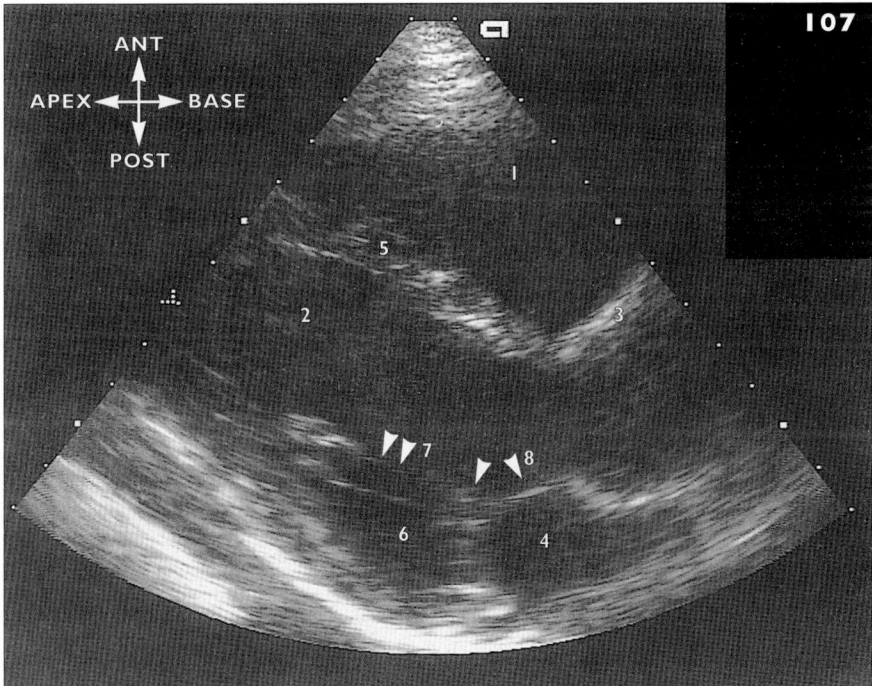

# 107:Answers

**107**  **1:** right ventricle. **2:** left ventricle. **3:** aorta. **4:** left atrium. **5:** ventricular septum. **6:** left ventricular posterior wall. **7:** chordae tendineae emerging from a papillary muscle in left ventricle. **8:** anterior and posterior leaflets of mitral valve.

The image (**107**) was obtained during ventricular systole because the mitral valve is closed and the aortic valve is open.

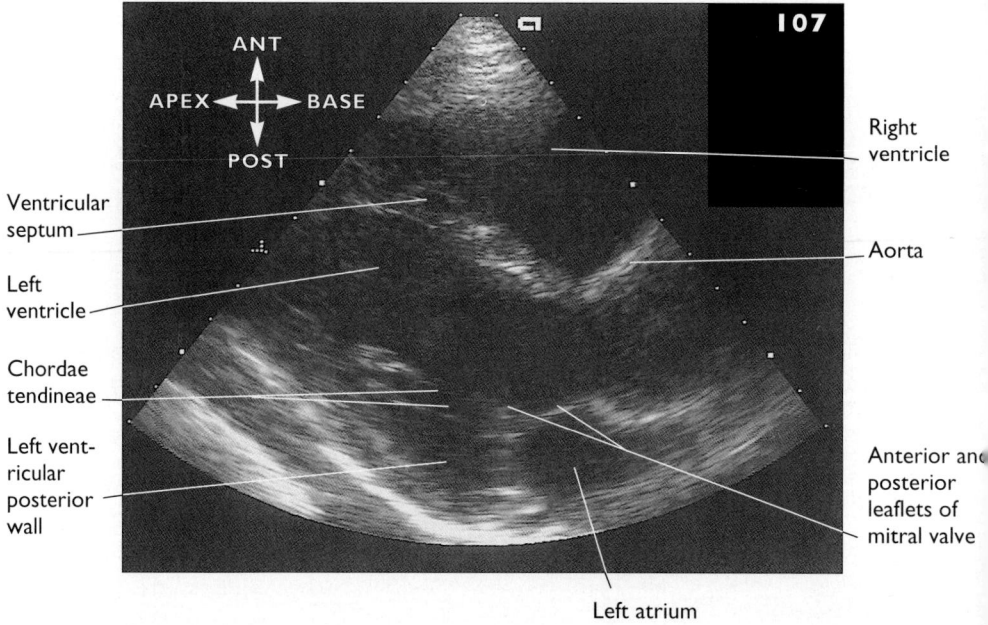

108 A 55-year-old woman presents to clinic complaining of recurring central, crushing chest pain. This pain radiates to her left arm and comes on with exercise. She is diagnosed as having angina pectoris. The pain of angina indicates that there is narrowing of the coronary arteries, which can be demonstrated radiologically by injecting contrast medium into the coronary arteries via a catheter (108a; arrow indicates stenosis in the anterior interventricular branch of the left coronary artery), usually inserted into the femoral artery in the femoral triangle. List the vessels through which the catheter must pass to reach the coronary arteries.

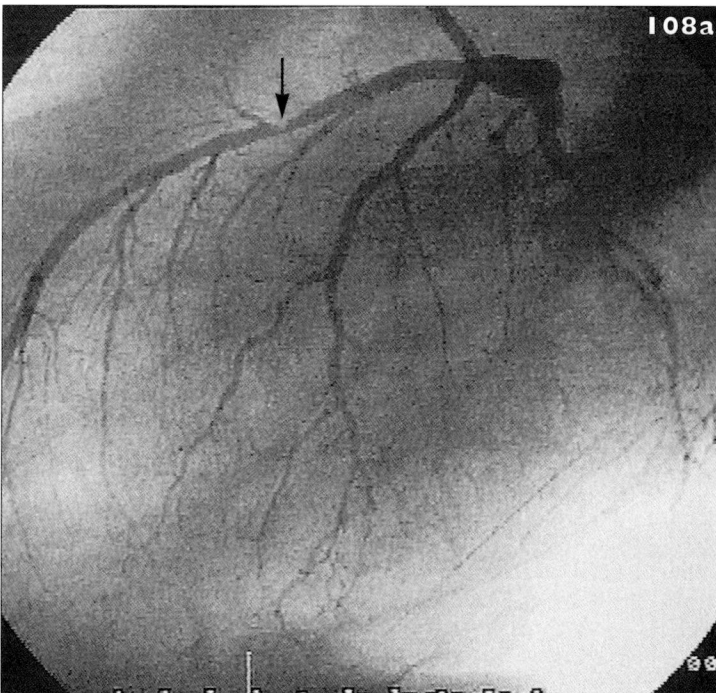

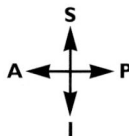

**108** From the femoral artery, the catheter would pass upwards into the external iliac and common iliac arteries and thence into the abdominal aorta (**108b**). Continuing upwards, the catheter would run through the thoracic aorta, arch of aorta, and ascending aorta, to come into the region of the aortic semilunar valve. Within this valve lie the openings for the right and left coronary arteries.

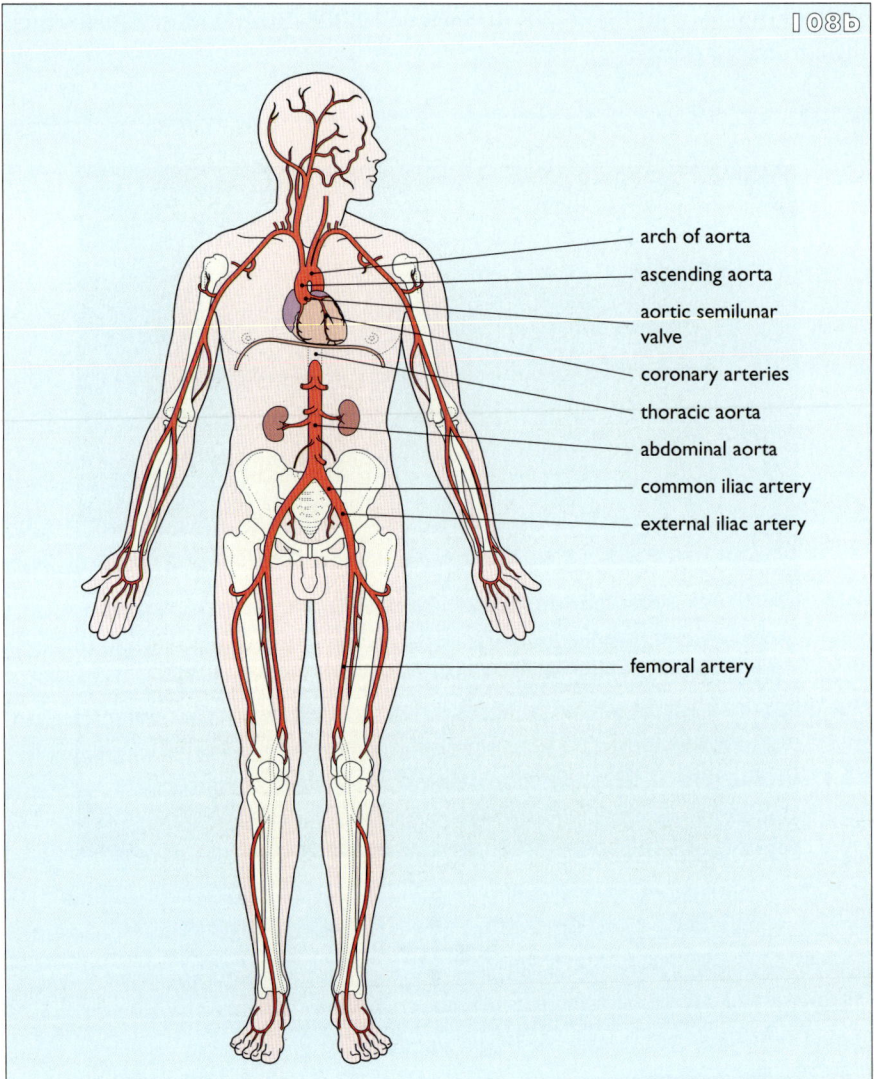

108b

arch of aorta

ascending aorta

aortic semilunar valve

coronary arteries

thoracic aorta

abdominal aorta

common iliac artery

external iliac artery

femoral artery

109 A 25-year-old man goes to his doctor with severe dyspnoea (difficulty in breathing) and using accessory muscles of inspiration (**109a**).
**i.** Give examples of accessory muscles of inspiration used (**a**) when the head is fixed or the neck extended, (**b**) when the arms are fixed or abducted at the shoulder.
**ii.** Where in the thoracic skeleton would you find (**a**) an anteriorly placed synovial joint, (**b**) a posteriorly placed synovial joint, (**c**) a primary cartilaginous joint (synchondrosis), (**d**) a secondary cartilaginous joint (symphysis)?

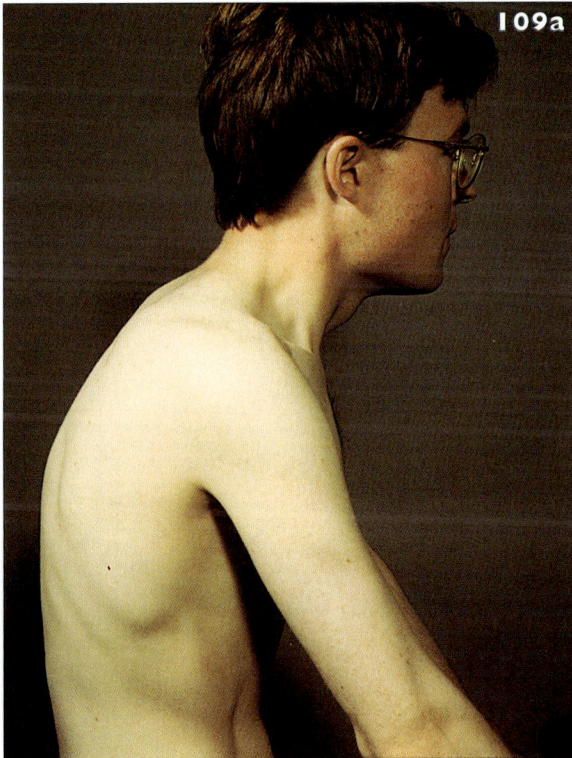

**109**   **i. (a)** Scalenus anterior, sternocleidomastoid, levator scapulae (**109b**); **(b)** pectoral, latissimus dorsi, serratus anterior (**109b**).
**ii. (a)** Sternoclavicular joint, chondrosternal joints (except first); **(b)** zygapophyseal joints, between heads or tubercles of ribs and vertebrae (costovertebral joints); **(c)** first chondrosternal joint; **(d)** manubriosternal joint.

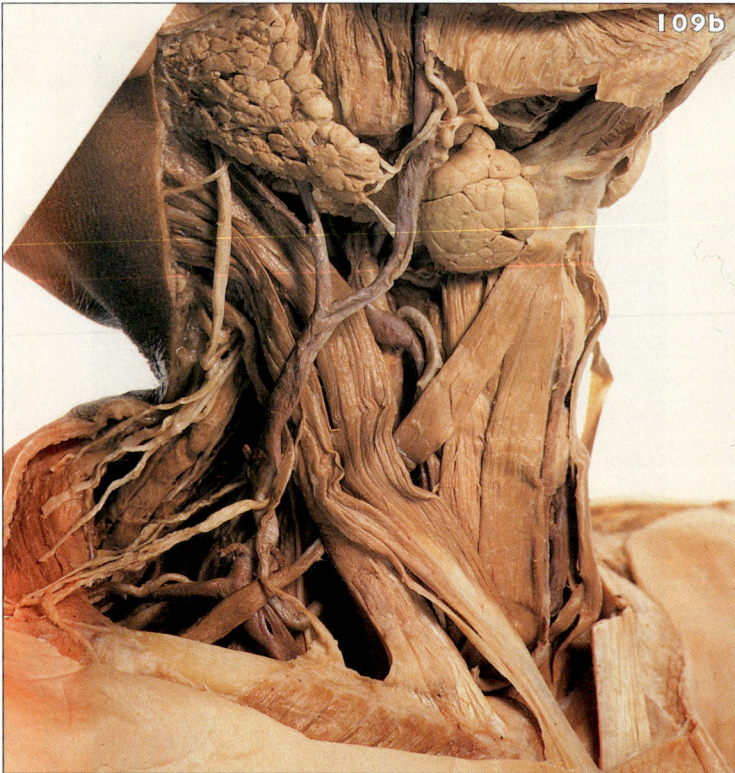

# THE ALIMENTARY AND UROGENITAL SYSTEMS

## Abdomen and Pelvis

For each of questions 110–129, show whether the statement provided is true or false.

110 Each renal medullary papilla discharges urine into a single major calyx.

111 The order of major structures in the renal sinus is ureter, renal artery, renal vein, from posterior to anterior.

112 The narrowest part of the male urethra lies within the prostate gland.

113 The pregnant uterus at term may be approached surgically through the anterior abdominal wall without passing through the peritoneal cavity.

114 The fibres of the external oblique muscle lie parallel to those of the internal oblique.

115 The inguinal lymph nodes drain the abdominal wall distal to the umbilicus.

116 The caudate lobe of the liver projects down into the upper recess of the lesser sac of the peritoneum (omental bursa).

117 The first part of the duodenum is directly related to the body of the gallbladder.

118 The transverse mesocolon is attached to the posterior abdominal wall along the upper edge of the pancreas.

119 The peritoneum lining the anterior abdominal wall is firmly attached to the transversalis fascia.

# 110–119: Answers

110  FALSE – each papilla discharges into a minor calyx.

111  TRUE – remember that the left renal vein crosses anterior to the aorta.

112  FALSE – the narrowest part is the external urethral orifice: the next narrowest is the membranous urethra. Urinary calculi may be trapped in these narrow areas.

113  TRUE – like the enlarged (full) bladder, the pregnant uterus 'dissects' the peritoneum off the anterior abdominal wall.

114  FALSE – the fibres run at right angles to each other. This is the basis of the 'grid iron' appendicectomy incision.

115  TRUE – look up the drainage of upper abdominal wall lymph.

116  TRUE – a median section of the abdomen passes through the caudate lobe, and shows it to be covered with peritoneum on both its anterior and posterior surfaces.

117  TRUE – the fundus of the gallbladder lies more anteriorly, and is usually related to the transverse colon.

118  FALSE – the body of the pancreas is prismatic, with three borders – superior, anterior, and inferior. The transverse mesocolon is attached along the anterior border.

119  FALSE – a layer of extraperitoneal fat intervenes.

120 The greater omentum is found between the stomach and the liver.

121 All parts of the small intestine are mobile within the peritoneal cavity.

122 Secretions of the seminal vesicle reach the urethra via the ejaculatory duct.

123 The sigmoid colon has no mesentery.

124 Peyer's patches of lymphoid tissue are found only in the walls of the large intestine.

125 The superior rectal artery is a branch of the inferior mesenteric artery.

126 The perineal membrane (urogenital diaphragm) lies in the transverse plane in the erect patient.

127 The proximal end of the fetal umbilical artery is represented in the adult by the obturator artery.

128 The ovary lies anterior to the broad ligament.

129 The peritoneal cavity of the female is, at least potentially, in communication with the exterior.

# 120–129: Answers

120 FALSE – the greater omentum is attached to the greater curve of the stomach and passes distally to overlie the transverse colon and the coils of small intestine. The lesser omentum runs from the lesser curvature of the stomach to the liver.

121 FALSE – all parts of the small intestine lie within the abdominal cavity, but almost all of the duodenum lies behind the peritoneum and therefore outside the peritoneal cavity.

122 TRUE – the duct of the vesicle joins with the vas (ductus) deferens to form the ejaculatory duct, which empties into the prostatic urethra.

123 FALSE – the sigmoid, or pelvic, colon has a very variable mesentery or meso-colon: sometimes this part of the colon is quite long, and can become twisted on its mesentery (sigmoid volvulus).

124 FALSE – Peyer's patches (aggregated lymphoid follicles) are found in the small intestine, most prominently in the ileum. They form part of the GALT (gut-associated lymphoid tissue) of the immune system.

125 TRUE – the middle rectal artery comes from the internal iliac artery, the inferior rectal from the (internal) pudendal.

126 TRUE – remember the position of the pelvis in the erect (standing) patient: the pubic tubercle and the anterior superior iliac spine lie in the same coronal plane.

127 FALSE – the proximal part of the umbilical artery is the superior vesical artery of the adult.

128 FALSE – the ovary lies behind the broad ligament.

129 TRUE – the communication is via the uterine (fallopian) tubes, uterus, and vagina. There is no such communication in the male, in whom the peritoneal 'cavity' is a closed one.

130 Identify the labelled structures **A–K** on the photograph (130) of the abdomen with the viscera *in situ*.

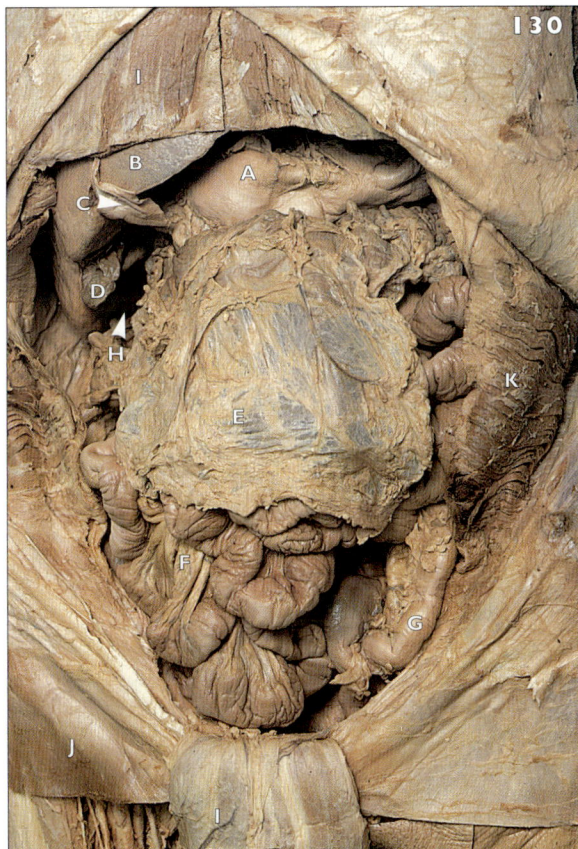

**130**  A: stomach. B: liver. C: falciform ligament; the ligamentum teres in the free edge of this structure is the obliterated remains of the umbilical vein of the foetus. D: fundus of the gallbladder. E: greater omentum; its attachment to the greater curvature of the stomach has been partly removed in this dissection. F: mesentery of the small gut. G: sigmoid colon, distinguishable as large bowel by the taeniae coli. H: in the depths of this dark hole is the hepatorenal pouch (of Rutherford–Morison). I: rectus abdominis, seen *in situ* but cut in the upper part, and reflected in the lower part. J: the aponeurosis of the external oblique, turned downwards and laterally. K: internal oblique (it is difficult to distinguish internal oblique from transversus abdominis in this photograph: the fibres of internal oblique pass upwards and medially in its upper part, whereas its lower fibres are more horizontal; it is attached to the lateral third of the inguinal ligament, whereas the transversus abdominis is attached to the lateral two-thirds).

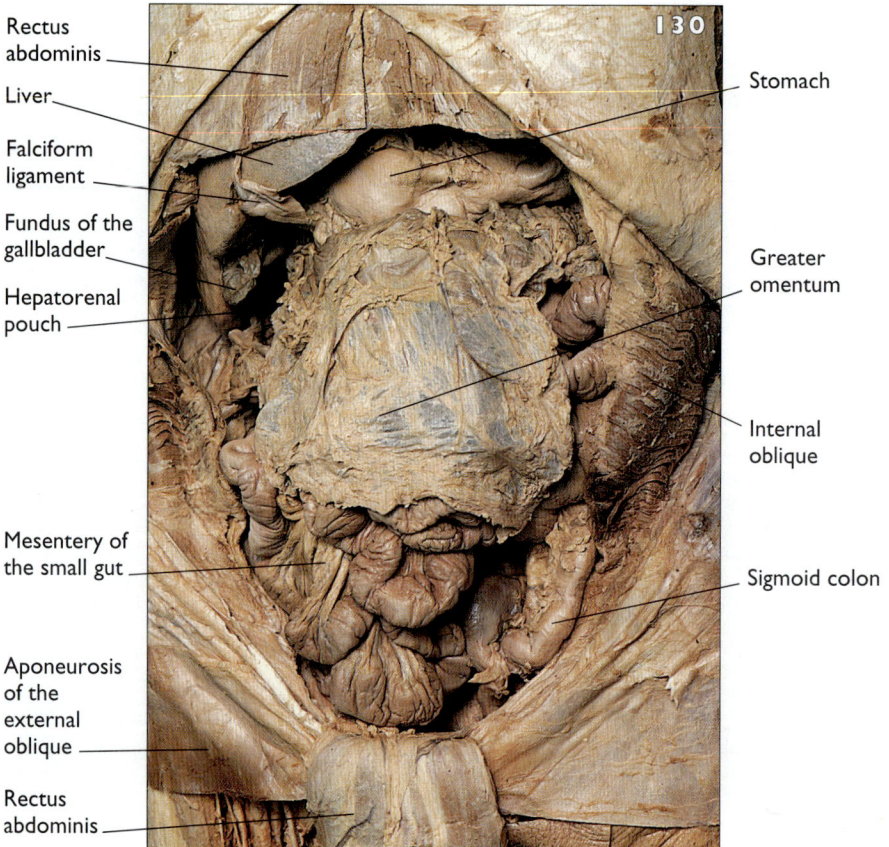

131 Identify the labelled structures **A–K** on the photograph (**131**) of the posterior abdominal wall.

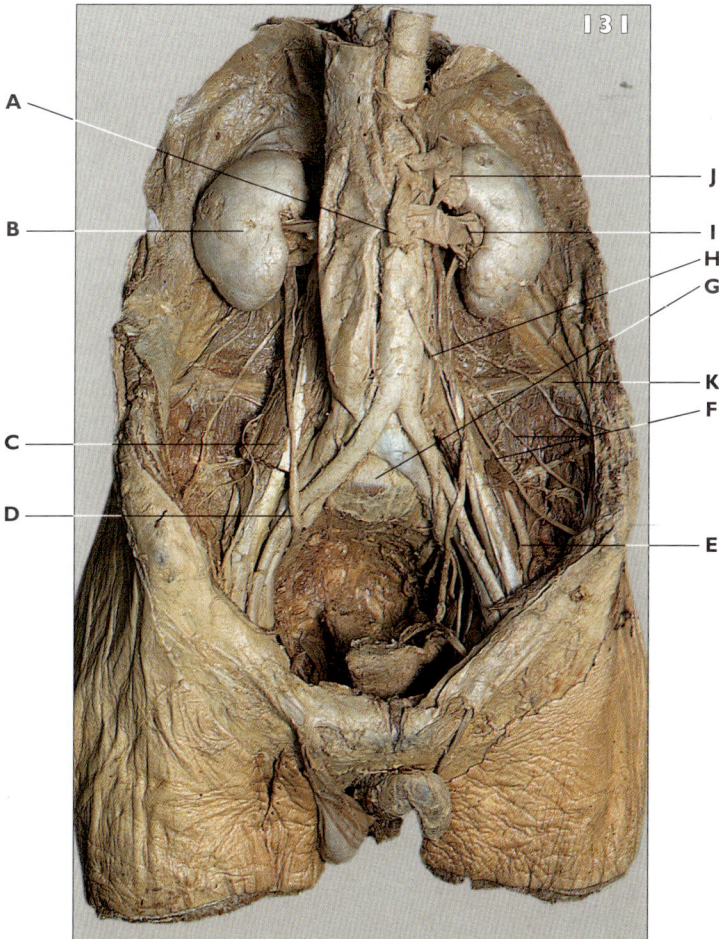

**131** **A:** superior mesenteric artery; in the intact abdomen, this part of the artery would be passing behind the neck of the pancreas and then in front of the uncinate process. **B:** right kidney. **C:** psoas major muscle. **D:** the ureter crossing the bifurcation of the common iliac artery. **E:** femoral nerve, which emerges from the lateral border of psoas major on its way downwards to pass behind the inguinal ligament. **F:** from above downwards, the iliohypogastric, ilio-inguinal and lateral femoral cutaneous nerves; the genitofemoral nerve can also be seen running on the anterior surface of psoas major on the left; the subcostal nerve is not visible. **G:** lumbosacral intervertebral disc. **H:** inferior mesenteric artery. **I:** left renal vein. **J:** left adrenal gland. **K:** posterior part of the iliac crest.

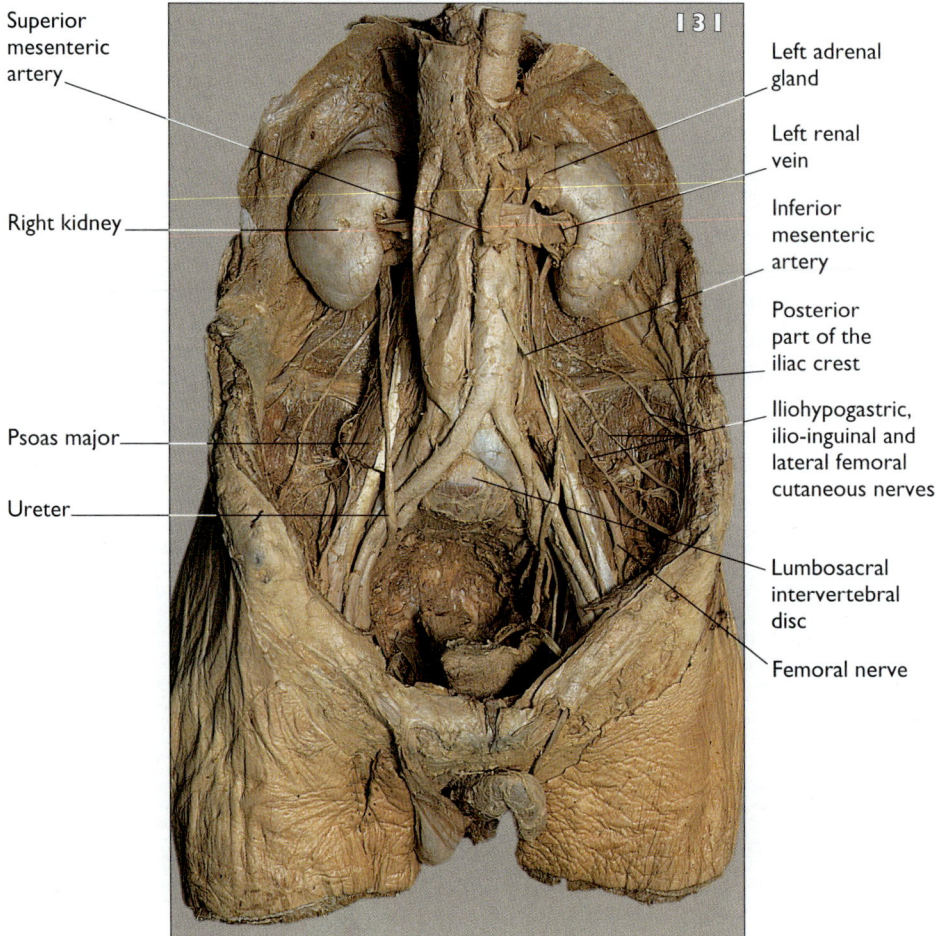

132 Shown below (132) is a sagittal section though the trunk, to the right of the midline. Identify the structures labelled A–G.

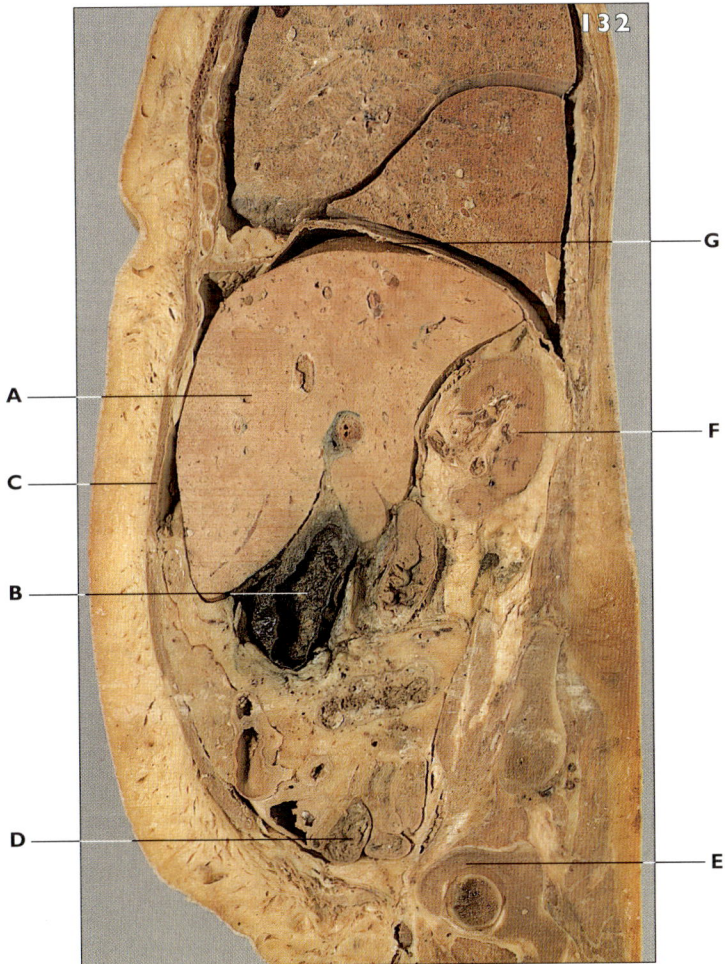

**132** A: liver. B: gallbladder. C: rectus abdominis (if the section were further to the right, the external and internal oblique and transversus abdominis muscles would be in this position; note also that the fundus of the gallbladder [see B] lies behind the lateral margin of rectus abdominis). D: pelvic colon. E: acetabular part of the hip bone. F: kidney. G: diaphragm.

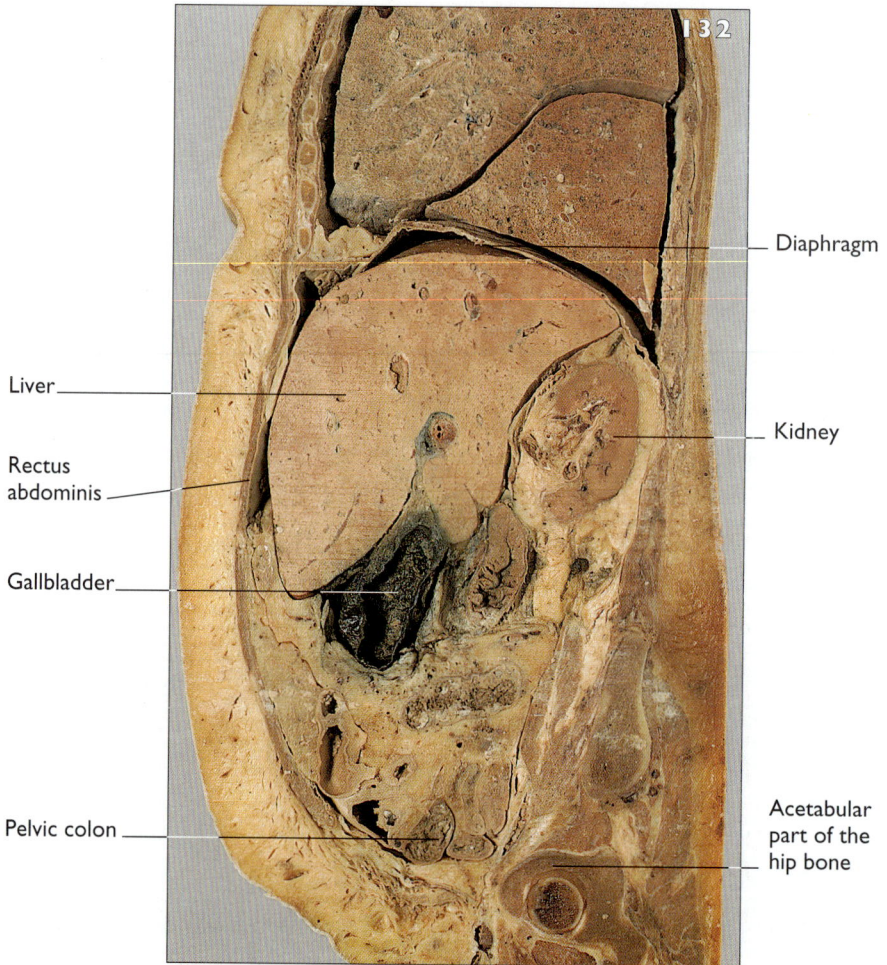

133 Shown below (133) is a coronal section through the trunk. Identify the features labelled **A–H**.

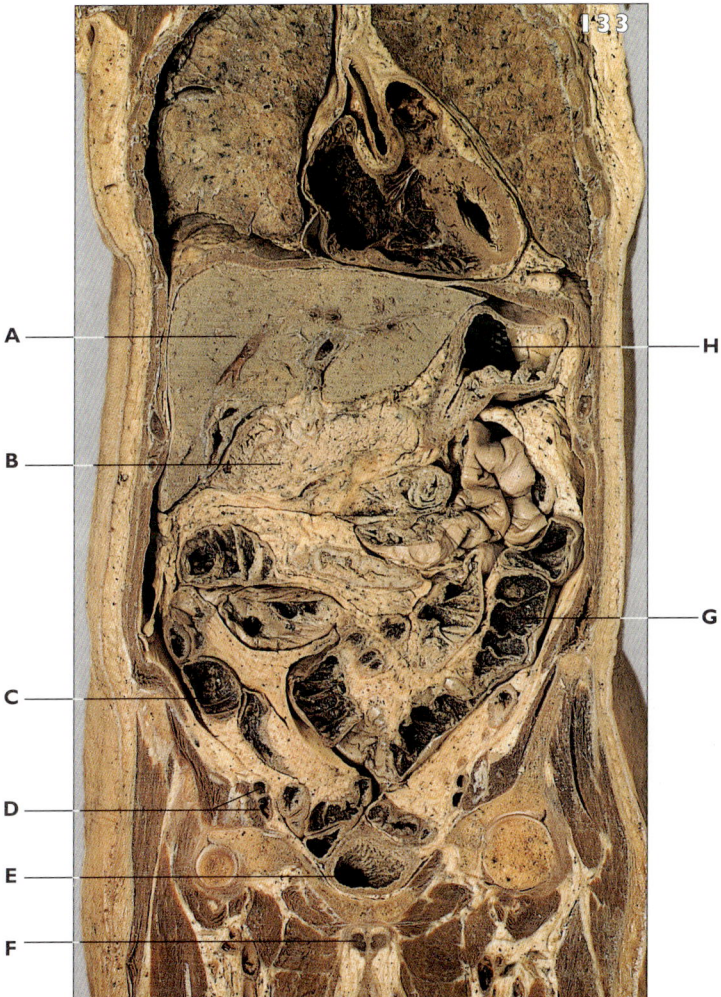

**133** **A**: liver. **B**: pancreas. **C**: caecum (note that it is difficult to differentiate the parts of the intestine in sections of the abdomen, but the position of this part in the iliac fossa strongly suggests that it is caecum). **D**: external iliac vessels, running anteriorly and downwards around the brim of the pelvis to pass beneath the inguinal ligament. **E**: bladder. **F**: crus of the clitoris. **G**: descending colon. **H**: stomach.

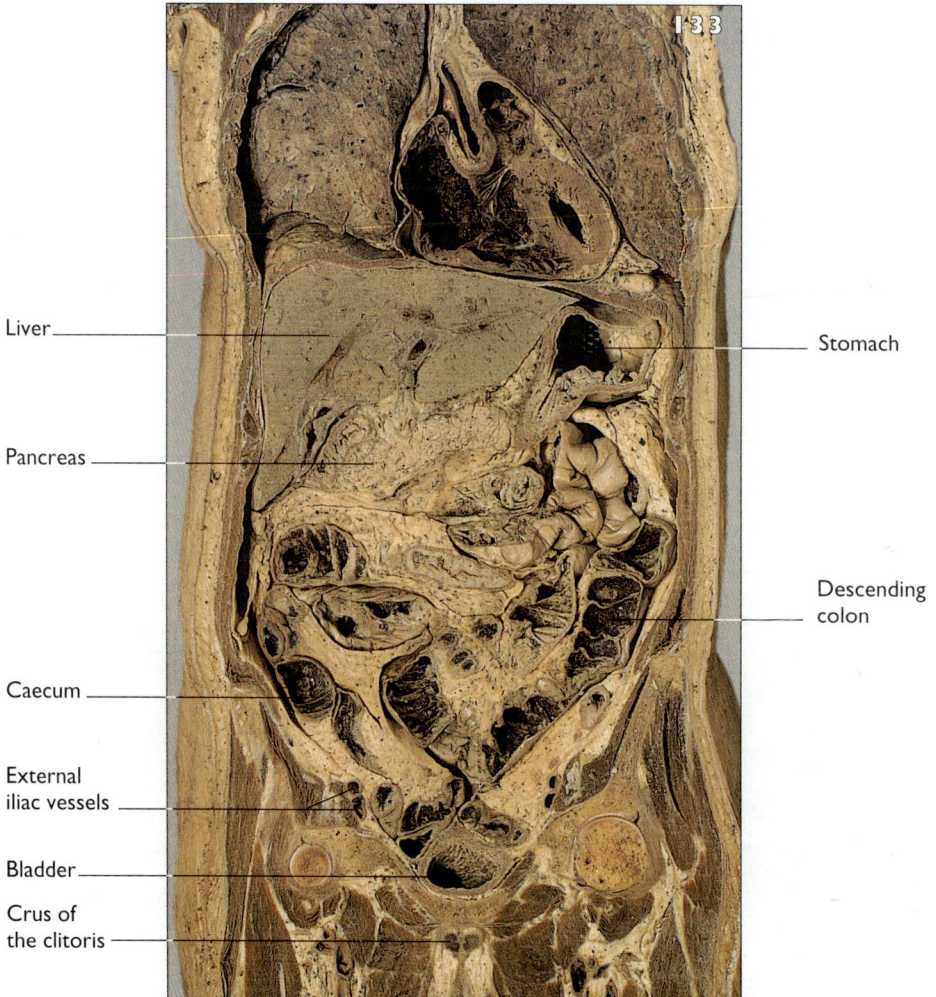

**134** Shown below (134) is the right hemipelvis of a female; part of the peritoneum covering the lateral wall of the pelvis has been pulled medially and backwards at X. Identify the structures marked **A–J**.

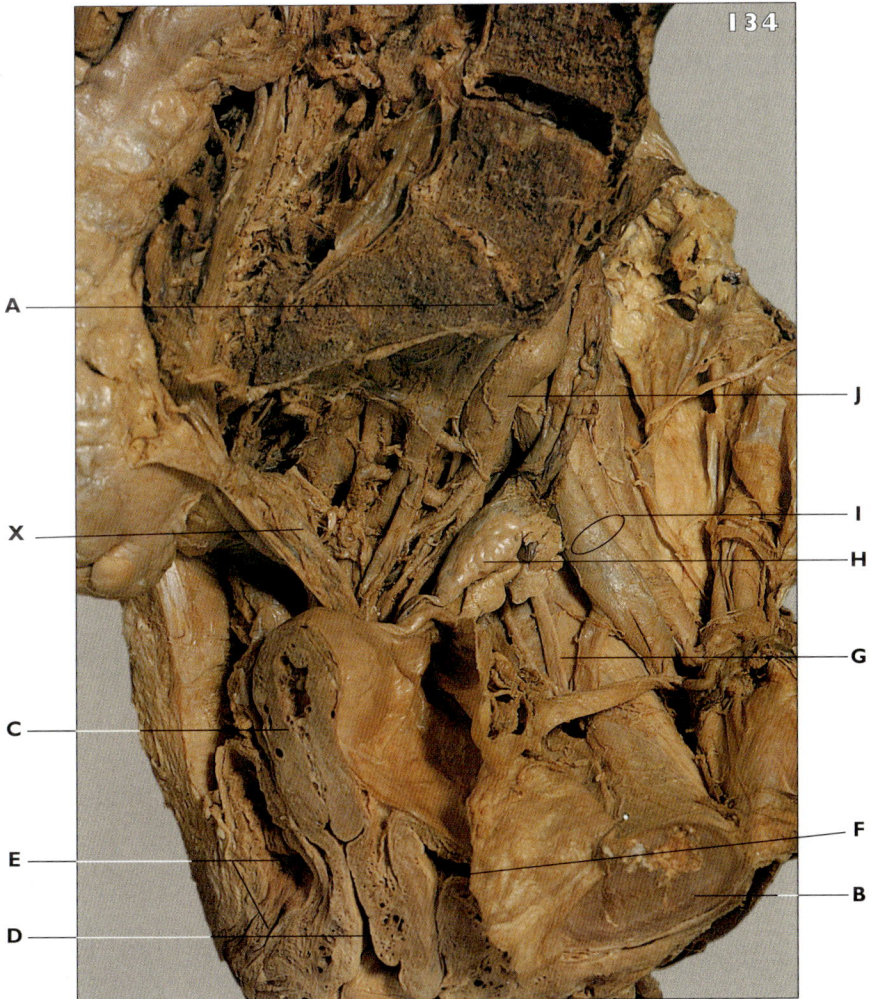

**134**   **A:** sacral promontory. **B:** pubic symphysis. **C:** uterus. **D:** vagina. **E:** rectum and anal canal. **F:** bladder. **G:** obturator nerve. **H:** ovary. **I:** external iliac vessels. **J:** internal iliac artery.

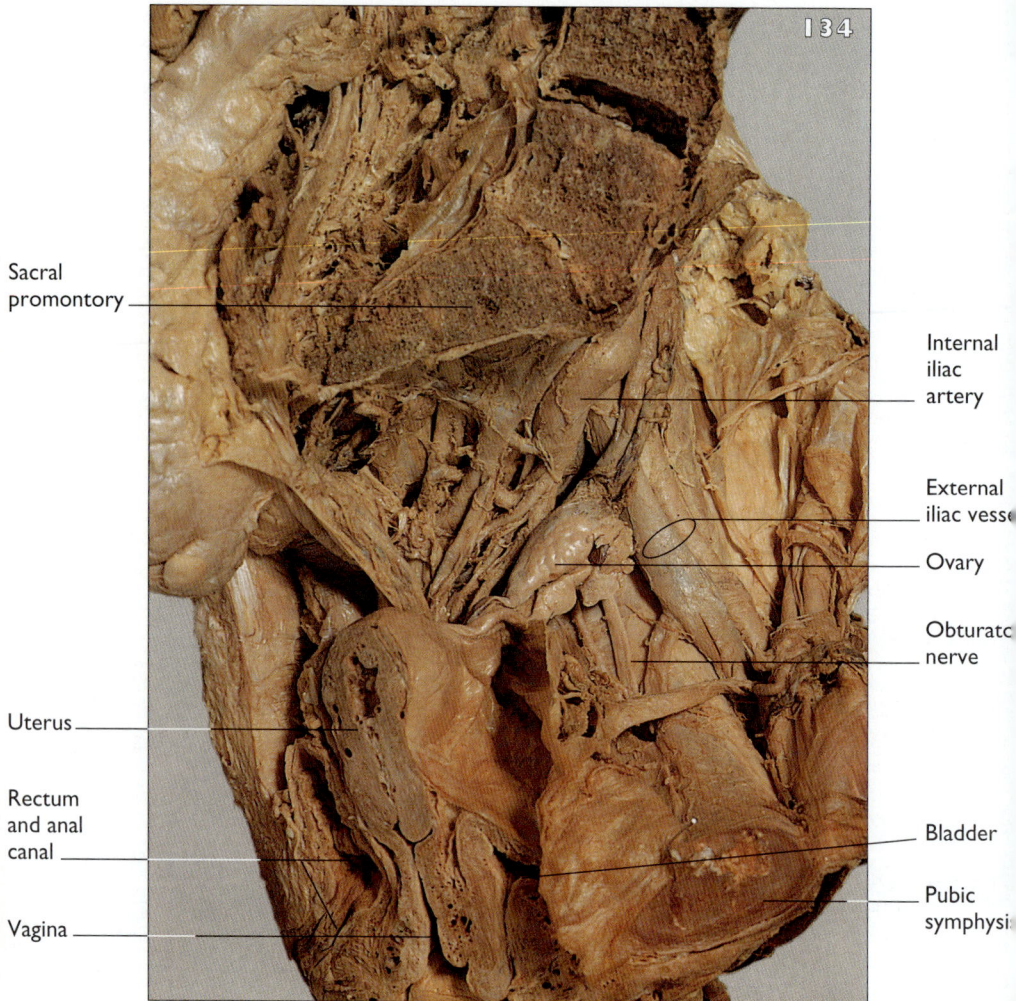

Sacral promontory

Internal iliac artery

External iliac vesse

Ovary

Obturato nerve

Uterus

Rectum and anal canal

Bladder

Vagina

Pubic symphysi

135  A sagittal section of a male pelvis (**135**). Identify the structures marked **A–H**.

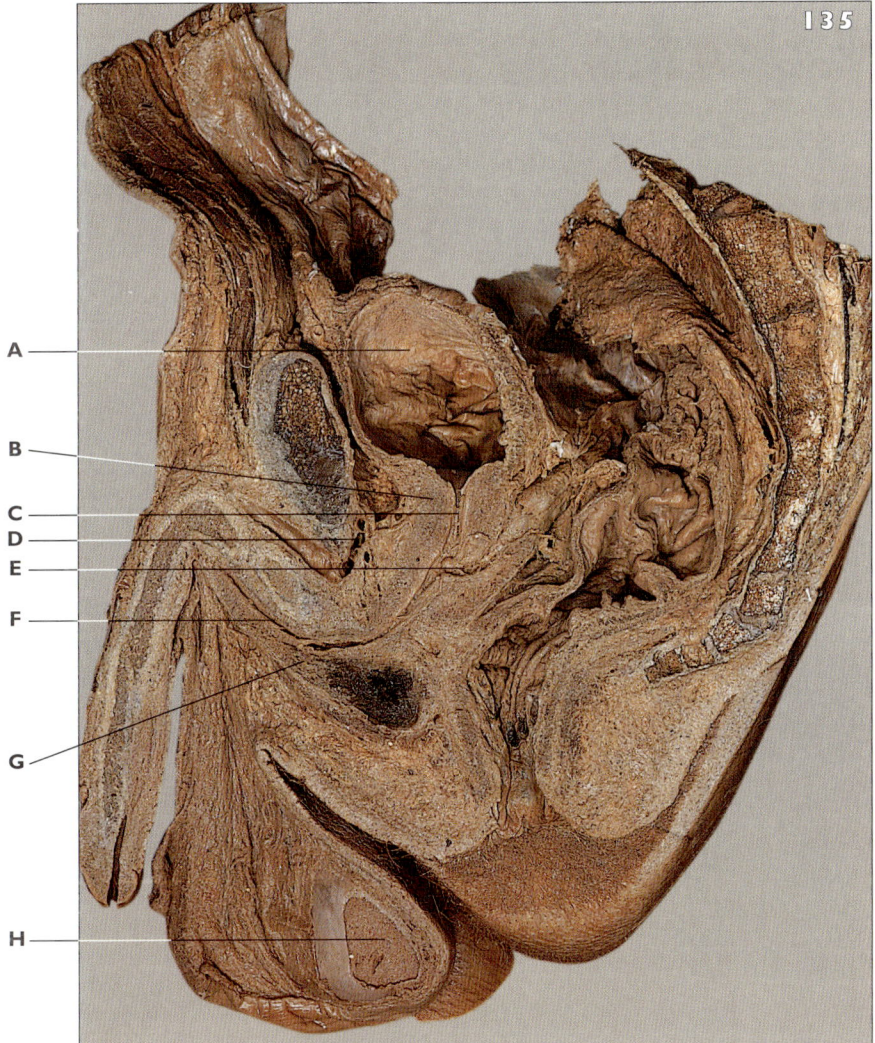

135  A: bladder. B: prostate. C: prostatic urethra. D: prostatic venous plexus.
E: ejaculatory duct. F: penile urethra. G: corpus spongiosum, or bulb of the penis.
H: testis.

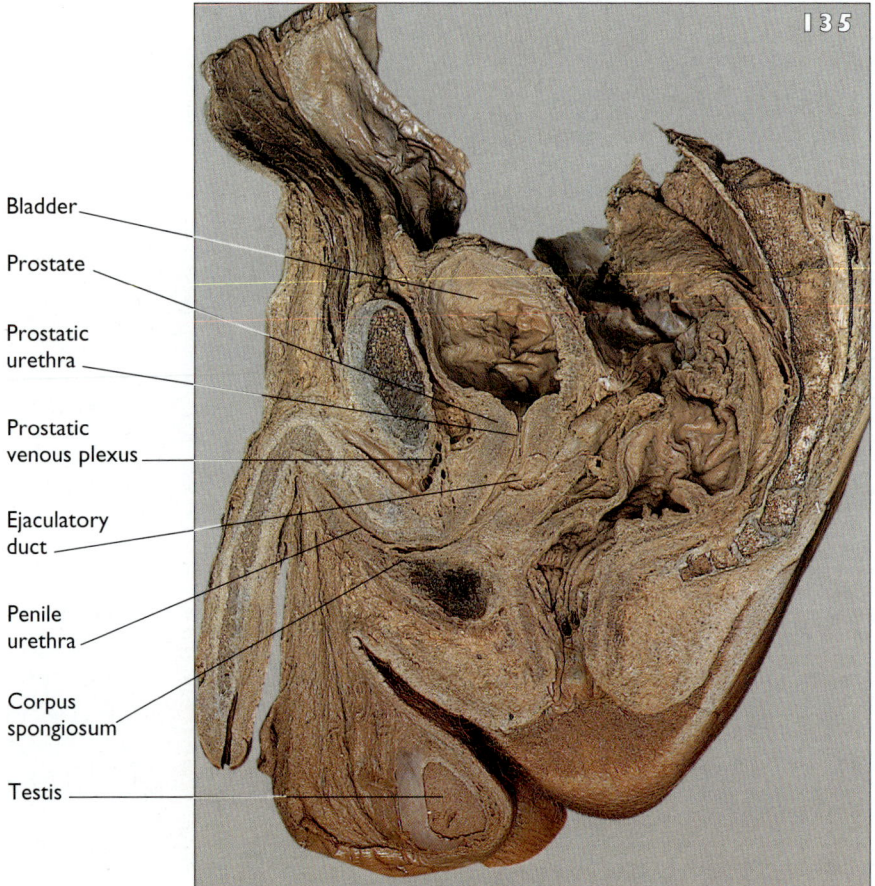

Bladder

Prostate

Prostatic
urethra

Prostatic
venous plexus

Ejaculatory
duct

Penile
urethra

Corpus
spongiosum

Testis

**136** A 46-year-old man complains of a dull, 'dragging' pain in his left groin, which has persisted for some months. When he is examined supine (lying down face upwards), no abnormality is found, but when he stands a soft swelling is seen and felt at the medial end of the groin (**136**). When he coughs, the swelling increases in size and extends into the upper part of the scrotum. A diagnosis of indirect inguinal hernia is made.

**i.** How much of the inguinal canal is traversed by an indirect inguinal hernia?

**ii.** Which structure lies immediately medial to the deep inguinal ring?

**iii.** Which tissue layer covering the spermatic cord corresponds to the internal oblique muscular layer of the abdominal wall?

**iv.** How would you identify the vas (ductus) deferens clinically in the living subject?

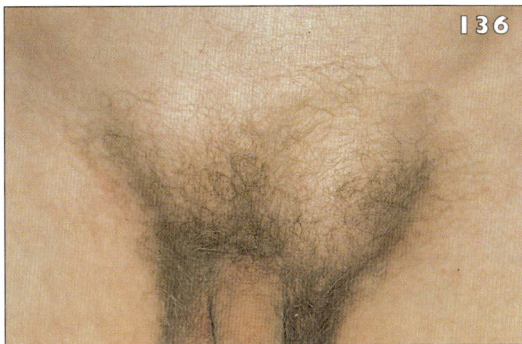

**137** A 70-year-old man presents with the clinical features of intestinal obstruction (abdominal pain and distension, vomiting, increased bowel sounds). Careful examination reveals a small, tense, tender mass in the right groin (**137**). The diagnosis of strangulated femoral hernia is made.

**i.** Through which opening does a femoral hernia leave the abdomen?

**ii.** What is the relationship of this opening to the pubic tubercle?

**iii.** What relation does the femoral canal bear to the inguinal ligament?

**iv.** What structure lies immediately lateral to the neck of a femoral hernial sac?

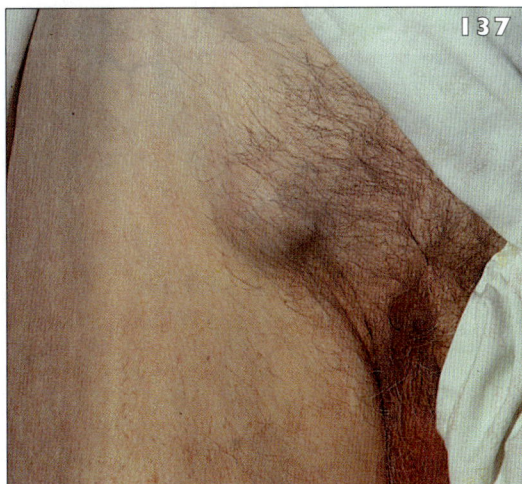

**136 i.** All of it. An indirect inguinal hernia passes through the deep inguinal ring, down the inguinal canal, through the superficial inguinal ring and into the scrotum. A direct inguinal hernia bursts through the posterior wall of the inguinal canal and then emerges through the superficial ring.

**ii.** The inferior epigastric artery. As there is no other strong structure in this position, some people describe the artery as forming one of the boundaries of the ring. This has become an even more important landmark now that many herniae are repaired laparoscopically.

**iii.** The cremasteric muscle and fascia.

**iv.** By palpation. It is a thin, hard structure, described as feeling like whipcord.

**137 i.** Through the femoral ring into the femoral canal.

**ii.** The femoral ring is bounded by the inguinal ligament in front, the lacunar part of the ligament medially. Since the inguinal ligament is attached to the pubic tubercle the femoral ring must be lateral to the tubercle. Anything passing through the ring from the abdomen will appear below and lateral to the pubic tubercle.

**iii.** The canal passes behind the ligament, but most of its length is inferior to the ligament.

**iv.** The femoral vein.

**138** A 67-year-old man has been unwell and losing weight for several months. He now presents to his doctor with jaundice and back pain, with some swelling of the lower limbs. On examination, a hard, immobile mass is felt in the upper abdomen, and eventually a diagnosis of carcinoma of the pancreas is made.

i. Explain anatomically (a) the jaundice, (b) the back pain, (c) the swelling of the lower limbs.

ii. Which peritoneal compartment lies between the pancreas and the posterior wall of the stomach?

iii. What is the relationship of the superior mesenteric artery to the pancreas?

**139** A 10-year-old boy is knocked off his bicycle and is found to have posterior fractures of the ninth, tenth, and eleventh ribs on the left. Soon after admission to hospital his condition deteriorates rapidly and he develops the signs of severe internal blood loss. An emergency surgical operation is necessary and an abdominal organ is removed (**139**).

i. What is the probable diagnosis, and what is the remedial surgical operation?

ii. In which peritoneal compartment does the injured organ lie?

iii. How is this organ supplied with blood?

iv. When this blood supply is being exposed and identified prior to surgical ligature, which major retroperitoneal organ is liable to be damaged?

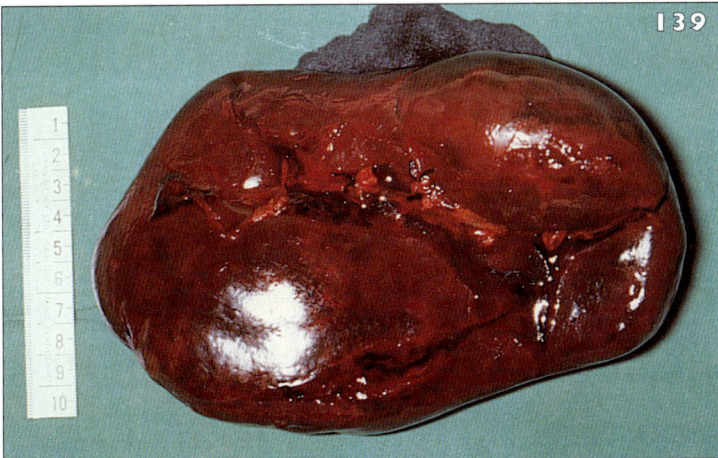

**138**  The illustration (**138**) shows a postmortem specimen of a carcinoma of the pancreas.

**i.** (a) Obstruction of the common bile duct.

(b) direct pressure on, or spread to, retroperitoneal soft tissues or the spine itself.

(c) obstruction of the inferior vena cava.

**ii.** The lesser sac (omental bursa).

**iii.** It passes behind the neck of the pancreas, then emerges to run anterior to the uncinate process.

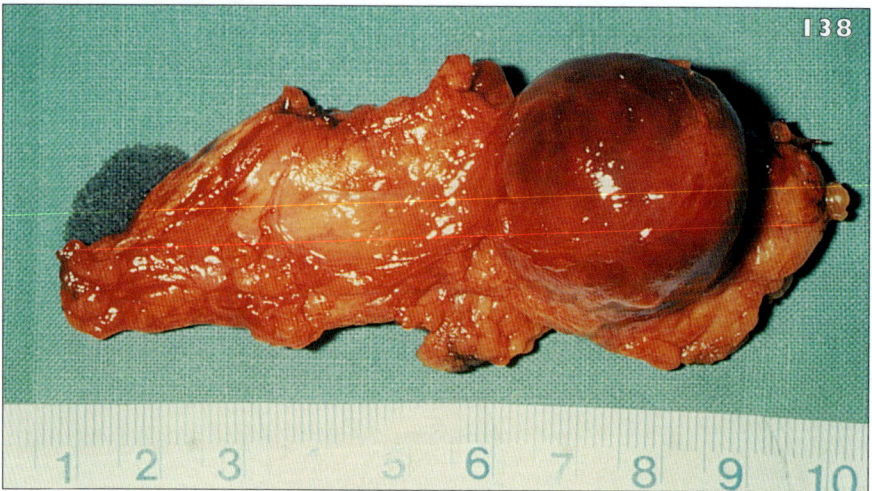

**139**  **i.** Ruptured spleen; splenectomy.

**ii.** In the greater sac of the peritoneum.

**iii.** Via the splenic artery from the coeliac trunk.

**iv.** The pancreas, whose tail lies in close relation to the splenic vascular pedicle. Damage to the pancreas could result in the formation of a pancreatic fistula.

**140** A 45-year-old overweight woman develops gallstones (**140**), and the resulting clinical problems lead her surgeon to advise cholecystectomy (removal of the gallbladder). At operation, the surgeon must take great care to identify accurately the arterial supply and the biliary drainage pattern of the gallbladder, in each patient.

**i.** Describe the route taken by the normal arterial supply to the gallbladder.

**ii.** Describe precisely the route taken by bile from the gallbladder into the intestinal tract.

**iii.** Describe the relationships of the major structures at the porta hepatis.

**iv.** Describe the innervation of the gallbladder.

**v.** What is the surface marking of the fundus of the gallbladder?

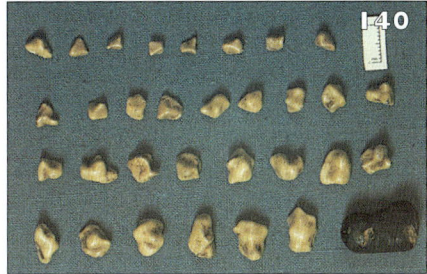

**141** A 12-year-old boy complains of central abdominal pain over a 12-hour period. The pain then worsens and moves to the right iliac fossa. There is tenderness, both abdominally (in the right iliac region) and rectally on the right. A clinical diagnosis of appendicitis is made and surgery is performed. When the abdomen is opened, the boy is found to have an inflamed pelvic appendix.

**i.** The diagram (**141**) shows the line of incision for a standard appendicectomy. What are the tissue layers that the surgeon must cut through or split in order to reach the peritoneal cavity?

**ii.** What is meant anatomically by a 'pelvic' appendix? In which other anatomical positions may the appendix lie?

**iii.** Where in the peritoneal cavity do pathological fluids such as pus and blood tend to collect (**a**) when the patient is sitting or standing, (**b**) when the patient is lying supine?

**iv.** Explain the terms visceral peritoneum and parietal peritoneum. How does the nerve supply differ?

**v.** To which dermatome is the pain of early appendicitis commonly referred?

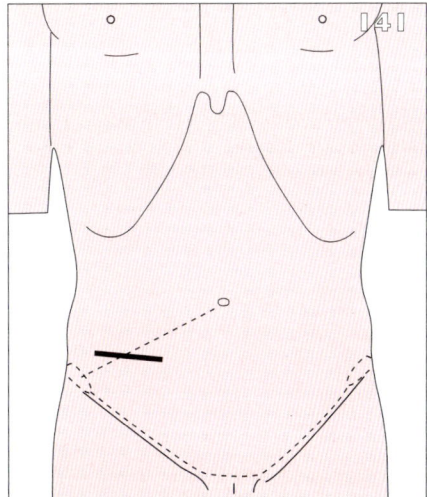

**140  i.** The cystic artery is usually a branch of the right branch of the hepatic artery proper. This latter is a terminal branch, with the gastroduodenal artery, of the comon hepatic artery, itself a branch of the coeliac trunk. Variation is common in this region and of great surgical importance: for example, the right, or even the common, hepatic artery may come from the superior mesenteric artery.

**ii.** Via the cystic duct, which joins the common hepatic duct to form the (common) bile duct or choledochus. This empties into the second part of the duodenum at the hepatopancreatic ampulla (of Vater). Again, variation is not unusual, but is not as common as variation in the arterial system.

**iii.** The (common) bile duct lies to the right of the hepatic artery proper; both lie anterior to the portal vein.

**iv.** Its nerve supply mainly consists of autonomic, sympathetic, and parasympathetic fibres reaching it from the coeliac plexus largely along the arteries described above. There is also a connection with the right phrenic nerve, via the coeliac and hepatic plexuses.

**v.** The tip of the ninth right costal cartilage, or the intersection of the linea semilunaris (lateral border of rectus abdominis) with the right costal margin.

**141  i.** Skin, superficial fascia, external oblique aponeurosis, internal oblique muscle, transversalis muscle, peritoneum, in that order.

**ii.** An appendix whose tip lies over the pelvic brim. The appendix may lie anterior or posterior to the caecum and terminal ileum. There is still controversy about the commonest position, despite numerous published surveys.

**iii.** (a) In the rectovesical or rectouterine pouch (of Douglas); (b) in the subphrenic space(s), in the hepatorenal pouch (of Rutherford–Morison), in the paracolic gutters, and in the pouch of Douglas.

**iv.** The visceral peritoneum is that which is closely applied to the viscera themselves, whereas the parietal lines the abdominal wall. The former, derived from splanchnopleure, has only an autonomic nerve supply like that of the viscera that it covers; the latter has a somatic supply corresponding to that of the particular area of body wall that it lines.

**v.** The pain is usually felt in the periumbilical region, said to be in the tenth thoracic dermatome. There is much dermatomal overlap.

**142** A 72-year-old man consulted his doctor, complaining of bleeding associated with defaecation. He reported that his father had died from 'cancer of the liver'. Investigation showed both a large polyp and a tumour in the sigmoid colon, which were surgically removed (**142**).

i. Why are polyps or tumours of the sigmoid colon more likely to present with rectal bleeding than polyps or tumours of the ascending colon?

ii. Cancer in the colon may spread via the veins or the lymphatics. What is the venous drainage of the sigmoid colon?

iii. Could the patient's father have had cancer of the colon, even though his death was described as due to 'cancer of the liver'?

iv. To which major lymph nodes does the colon first drain?

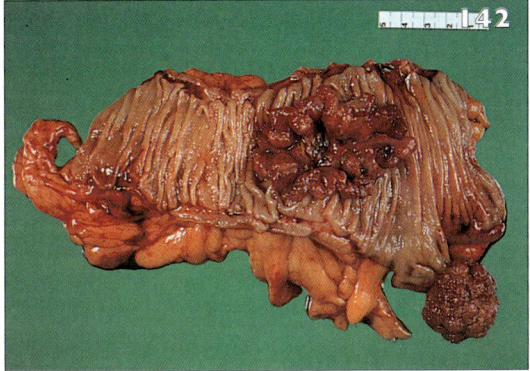

**143** A 40-year-old man had repeated episodes of colicky pain in the back, which radiated around the side of the abdomen towards the groin. A radiograph showed a small dense opacity close to the transverse process of L3. Figure **143** shows an IVP.

i. In which tubular, soft-tissue structure is the calcified mass likely to be?

ii. How does the structure obtain its blood supply?

iii. Why is the pain colicky in nature?

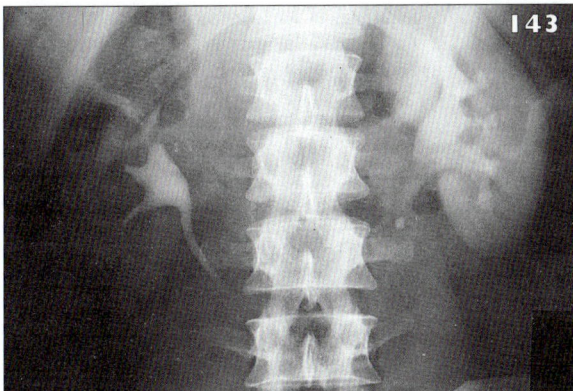

**142 i.** Because the sigmoid colon is nearer the anus and so the blood has less time to become changed or mixed with the contents of the large bowel.
**ii.** Via sigmoid veins to the inferior mesenteric vein. This then drains into the splenic vein and then into the portal vein.
**iii.** Yes. His father might have had cancer of the colon that metastasized (spread) to the liver.
**iv.** To the preaortic lymph nodes around the root of the inferior mesenteric artery. There are smaller nodes along the colonic border of the mesentery and along the branches of the inferior mesenteric artery.

**143 i.** The ureter.
**ii.** From nearby vessels along its length, namely renal, segmental branches of the aorta, common and internal iliac, gonadal, and vesical arteries.
**iii.** Because it arises from excessive contractions of the smooth muscle of the ureteric wall as it tries to propel the stone down to the bladder.

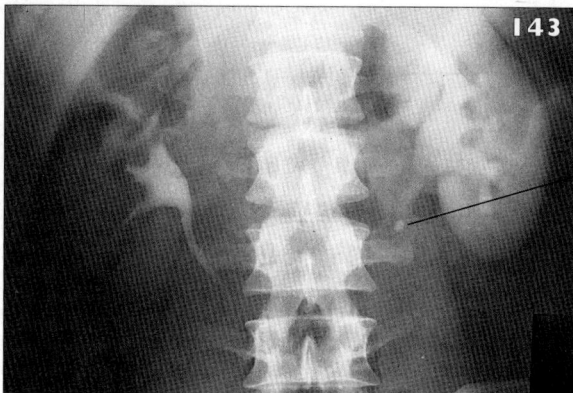

Site of stone

**144** A 70-year-old woman visited her doctor complaining of vague abdominal pain that had become worse recently. On examination, the doctor felt a large, pulsatile mass slightly to the left of the midline. He diagnosed an aneurysm (enlargement) of a large artery. The illustration (**144**) shows the aneurysm at operation.

i. Which artery was involved?
ii. If the aneurysm involved branches of the vessel, which branches might be affected?
iii. Would the gut be affected and, if so, which part, if these branches became blocked?

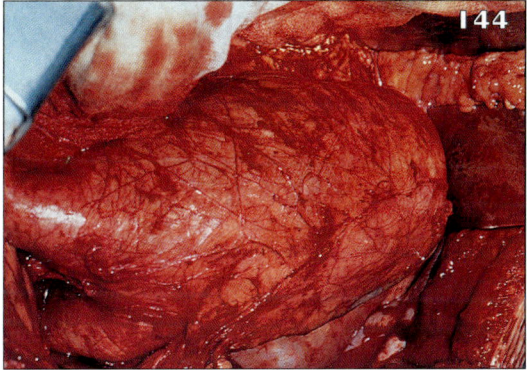

**145** A 60-year-old man complained of a lump in his scrotum. On examination, the lump was found to be attached to the front of the testis, partly covering it (**145**). The lump was fluctuant and transilluminated, allowing the doctor to diagnose a fluid-filled, cystic swelling.

i. What is the name of the structure that has become enlarged and fluid-filled in this case?
ii. If the swelling had been behind and above, but still attached to, the testis, which structure is most likely to have been involved?
iii. Tumours of the testis cause swelling of the testis itself, but often the shape and surface of the testis feel normal. Why?
iv. Tumours of the testis commonly spread via the lymphatics. To which group of lymph nodes do the lymphatics of (a) the testis and (b) the scrotum drain?

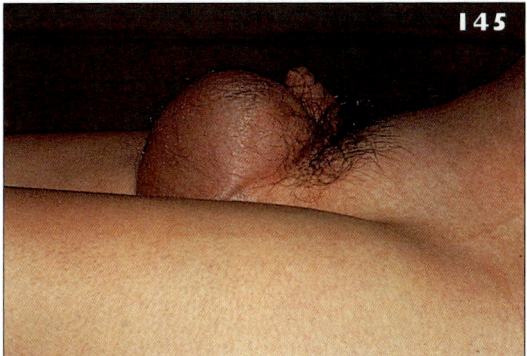

**144  i.** The abdominal aorta.
**ii.** Renal arteries, common iliac arteries, inferior mesenteric artery, lumbar arteries. The superior mesenteric and coeliac arteries are given off proximal to the renal arteries. Obstruction of the gonadal arteries would be the least of the patient's worries!
**iii.** Yes: the hindgut, distal to the transverse colon.

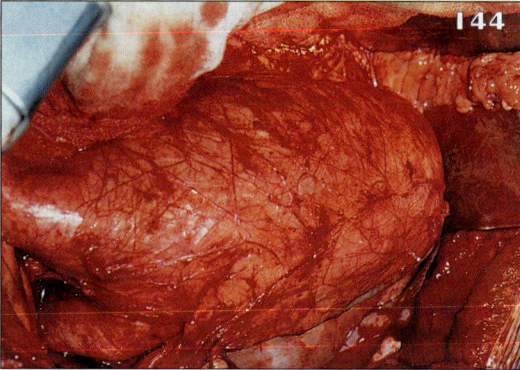

**145  i.** The tunica vaginalis of the testis.
**ii.** The epididymis.
**iii.** The dense fibrous tissue of the tunica albuginea prevents the examiner from feeling the surface of the tumour within the testis.
**iv.** (a) To the lateral and preaortic lymph nodes near the origin of the testicular artery.
(b) To the inguinal lymph nodes.

**146** A 50-year-old-woman suffering from heavy menstrual bleeding is diagnosed as having uterine fibroids. These are benign smooth-muscle tumours of the uterine wall. The fibroids are large enough to cause abdominal swelling.
i. Why does she also complain of frequency of micturition?
ii. Extensive fibroids are often treated by an operation (hysterectomy) to remove the whole uterus. What is the blood supply of the uterus?
iii. Which structure is particularly at risk when ligating and cutting the main arteries to the uterus?
iv. Some women with large fibroids suffer from swelling or varicose veins of the legs. Why?

**147** i. What is the surface marking of the highest point of the liver in quiet respiration?
ii. What is the surface marking of the fundus of the gallbladder?
iii. Is the bladder normally palpable?
iv. Which point is usually the site of greatest tenderness in appendicitis? Is this the surface marking of the root of the appendix?

**146** The illustration (**146**) shows the relationship of the uterine artery to the ureter in a dissected right half female pelvis.
**i.** Because the enlarged uterus is pressing on the bladder.
**ii.** Uterine arteries. Even though these anastomose with the ovarian arteries superiorly and the vaginal arteries inferiorly, the main supply is via the uterine arteries.
**iii.** The ureter.
**iv.** Because pressure from the enlarged uterus can obstruct the iliac veins.

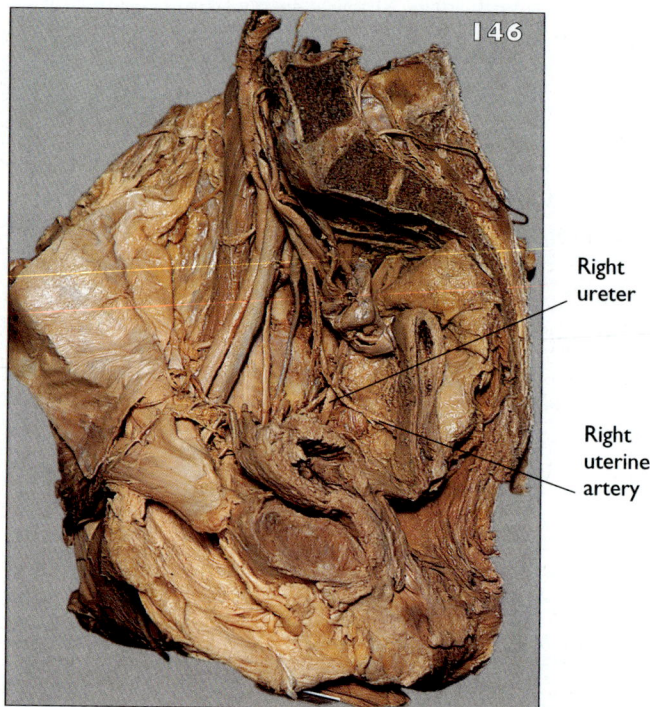

Right ureter

Right uterine artery

**147**  **i.** A point immediately below the diaphragm, reaching as high as the fifth costo-chondral junction on the right.
**ii.** A point between the tips of the right ninth and tenth costal cartilages, or the point where the outer border of rectus abdominis crosses the costal margin on the right.
**iii.** No; it is palpable only when distended with urine.
**iv.** McBurney's point, which is one-third of the way along a line from the anterior superior iliac spine to the umbilicus. No; the root of the appendix is generally considered to lie on the transtubercular plane, which is a little below McBurney's point; however, the difference is of no practical importance.

# THE HEAD AND NECK

For each of questions 148–167, show whether the statement provided is true or false.

148 The pituitary fossa is a central depression on the intracranial aspect of the ethmoid bone.

149 The flat bones of the calvaria are traversed by venous channels called the diploë.

150 The lateral and medial pterygoid muscles originate entirely from the lateral pterygoid plates.

151 The typical cervical vertebra differs from vertebrae in other regions in that its spine is bifid.

152 Damage to the spinal accessory nerve will impair the ability to shrug the shoulders.

153 The oculomotor nerve supplies the lateral rectus muscle.

**148** FALSE – the pituitary, or hypophyseal fossa, is a central depression on the superior surface of the body of the sphenoid bone. It lies within the middle cranial fossa.

**149** TRUE – diploic venous channels connect meningeal veins and dural venous sinuses with pericranial veins, thus providing a route for the spread of infection intracranially.

**150** FALSE – the lateral pterygoid muscle originates as two heads: the larger inferior head arises from the lateral surface of the lateral pterygoid plate of the sphenoid bone; the superior head takes origin from the infratemporal surface of the greater wing of the sphenoid bone in the roof of the infratemporal fossa. The medial pterygoid muscle also has two heads: the larger (deep) head arises from the medial surface of the lateral pterygoid plate; the superficial head originates from the maxillary tuberosity and the pyramidal process of the palatine bone.

**151** TRUE – the typical cervical vertebra also differs from other vertebrae in that the body is small, nearly cylindrical but flattened anteroposteriorly, the body shows lips (unci) at each side of the upper surface posteriorly, the transverse process is small and contains a foramen transversarium for the passage of vertebral blood vessels, and the vertebral foramen is triangular and larger than any other regions of the vertebral column. Remember that the first, second, and seventh cervical vertebrae have some unique features.

**152** TRUE – the spinal accessory nerve is derived from the upper five segments of the cervical spinal cord. They form a trunk which passes intracranially through the foramen magnum to join the cranial part of the accessory nerve. On leaving the cranium through the jugular foramen, the spinal and cranial parts separate. The spinal part of the accessory nerve supplies sternocleidomastoid and trapezius. It is thus through its innervation of the trapezius that damage to the spinal accessory nerve will impair the ability to shrug the shoulder. In addition, the loss of innervation to sternocleidomastoid will affect the ability to tip the head towards the shoulder, to rotate and direct the face towards the opposite side, and to move the head forwards.

**153** FALSE – lateral rectus is innervated by the abducens nerve (VI). The oculomotor nerve (III) supplies the superior, medial, and inferior recti, inferior oblique, levator palpebrae superioris and certain intraocular muscles (including constrictor pupillae).

154 The cavernous sinus links with the extracranial venous system through the ophthalmic veins.

155 All the infrahyoid muscles are innervated by branches directly from the ansa cervicalis.

156 The cervical plexus is formed by the posterior primary rami of the upper four cervical nerves.

157 The carotid sinus is innervated by the glossopharyngeal nerve.

158 The auriculotemporal nerve supplies the skin of the tragus and external auditory meatus.

159 The buccinator muscle is pierced by the parotid duct opposite the upper second molar tooth.

154  TRUE – the linkage of the cavernous sinus with the ophthalmic veins is impor
tant because of the possibility of intracranial spread of infection from the face.

155  FALSE – geniohyoid and thyrohyoid are innervated by fibres from the first
cervical segment running in the hypoglossal nerve (XII). The 'descendens
hypoglossi', which forms the superior root of the ansa cervicalis, has the same
segmental origin, but leaves the hypoglossal nerve lateral to the occipital artery
and runs anterior to the internal and common carotid arteries to join the lower
root. The superior head of omohyoid takes its innervation from a branch of the
descendens hypoglossi.

156  FALSE – the cervical plexus is formed by the anterior primary rami of the upper
four cervical nerves. The posterior primary roots of the first cervical segmental
nerve supply the rectus and obliquus capitis of the suboccipital triangle; the
posterior primary root of the second cervical nerve forms the greater occipital
nerve supplying the skin of the posterior occipital region. Muscular branches of
the posterior primary root of the second cervical nerve supply semispinalis
capitis, splenius, and longissimus capitus.

157  TRUE – the carotid sinus nerve is sensory to the carotid sinus and also to the carotid
body, which consists of glomeruli located within adventitial tissue in the carotid
bifurcation. The nerve is substantial enough to be dissected out and anaesthetized
locally during a carotid endarterectomy, so that cardiovascular centres within the
brain stem are not perturbed by afferent nerve impulses during surgery.

158  TRUE – the auriculotemporal nerve is also sensory to the tympanic membrane.
Additionally, it carries postganglionic parasympathetic secretomotor fibres from
the otic ganglion to the parotid gland. In mumps, the production of saliva is inter-
rupted owing to impairment of the secretomotor innervation, resulting in a dry mouth.

159  TRUE – the duct leaves the supero-anterior margin of the parotid and crosses
the masseter on the lateral side of the face before passing through
buccinator and into the mouth.

160 The facial artery crosses the inferior border of the mandible at the anterior margin of the masseter.

161 Inferior oblique moves the eye downwards and outwards.

162 The ciliary ganglion is associated with the secretomotor supply to the lacrimal gland.

163 The infratemporal fossa is bounded medially by the middle constrictor.

160     TRUE – in order to allow for movements of the mandible when the mouth is opened, the facial artery is tortuous as it crosses the border of the mandible. When the mouth is opened, the 'slack' in the length of the artery is taken up without the vessel being unduly stretched. There is, therefore, a certain amount of variation in the position of the facial artery as it crosses the mandibular border, which may make a facial artery pulse relatively difficult to locate.

161     FALSE – Inferior oblique arises from the medial part of the front of the orbital floor. It passes posterolaterally and is inserted into the eyeball behind its equator and above the level of the origin of the optic nerve (i.e. in the upper posterolateral quadrant of the eyeball). Contraction of inferior oblique alone would therefore result in the eye moving upwards and outwards. In reality, inferior oblique works in concert with superior rectus in such a way that the dual action of the muscles is to move the eye upwards.

162     FALSE – the ciliary ganglion contains the cell bodies of the postganglionic parasympathetic neurons that supply two intrinsic muscles of the eye – the constrictor pupillae in the iris and the ciliaris muscle attached to the lens. The preganglionic neurons of this efferent autonomic pathway are found in the Edinger–Westphal nucleus of the oculomotor nerve in the upper midbrain. Constriction of the pupil occurs either when the pathway is stimulated by an increase in the level of ambient light or if the dilatator pupillae muscle is paralyzed, as occurs in Horner's syndrome. During convergence of the eyes, when the gaze transfers from a distant to a near object, both constrictor pupillae and ciliaris muscles are active, so that the diameter of the pupil and the convexity of the lens can focus light onto the retina. As both oculomotor nuclei are connected across the midline in the pretectal region, this occurs in both eyes, even if one eye is covered (the consensual reflex). The secretomotor supply to the lacrimal gland is via the pterygopalatine ganglion.

163     FALSE – the medial boundary of the infratemporal fossa is the lateral plate of the pterygoid bone from which the lower head of the lateral pterygoid muscle arises. The superior constrictor muscle of the pharynx also bounds the infratemporal fossa (not the middle constrictor). Thus, infection that passes deep into the infratemporal fossa might spread into the pharyngeal tissue spaces.

164  The extrinsic muscles of the tongue alter the shape.

165  The pharyngeal nerve plexus is derived from the glossopharyngeal, vagus, and accessory nerves, plus sympathetic branches.

166  The vocal fold is the free upper border of the cricovocal membrane.

167  Within the middle meatus of the nose, the hiatus semilunaris receives the openings of the frontal, anterior ethmoidal, and maxillary air sinuses.

164 FALSE – the shape of the tongue is altered by its intrinsic musculature. Within the tongue there are longitudinal, transverse, and vertical bundles of muscle which are all innervated by the hypoglossal nerve. If the hypoglossal nerve is lesioned unilaterally, there is muscle wasting on the affected side and the tongue deviates towards that side, as the action of the intrinsic muscles on that side is unopposed. The extrinsic muscles of the tongue, all supplied by the hypoglossal nerve (with the exception of the palatoglossus muscle), alter the position of the tongue.

165 TRUE – the intrinsic muscles of the wall of the pharynx and the striated muscle of the upper oesophagus are supplied by nerve fibres that originate from the nucleus ambiguus in the medulla oblongata. These fibres, whether classified as cranial accessory or vagal, all enter the vagus nerve trunk from which a pharyngeal branch passes to the pharynx at the level of the middle constrictor. Here, the vagal branches mingle with branches of the glossopharyngeal nerve, which are sensory to the mucous membrane of the pharynx, and with sympathetic vasomotor nerves derived from the superior cervical ganglion, to form the pharyngeal plexus.

166 TRUE – the core of the vocal fold is a bundle of striated muscle fibres sometimes termed the vocalis muscle; this muscle is part of the thyroarytenoid, which lies lateral to the vocal cord and is responsible for pulling the arytenoid cartilage, thereby relaxing the vocal cord. With the exception of cricothyroid (supplied by external laryngeal nerve), all the intrinsic laryngeal muscles are supplied by the recurrent laryngeal branch of the vagus nerve.
Unilateral damage to the recurrent laryngeal nerve will result in hoarseness, as the vocalis on the affected side is paralyzed. Bilateral paralysis results in the closure of the rima glottidis, causing asphyxia and death.

167 TRUE – the opening of the maxillary air sinus is located on the upper part of the medial wall of the sinus and, therefore, if it becomes infected in sinusitis, postural drainage is required. This is in contrast to the frontal sinus, which is drained inferiorly.

**168** Identify the structures labelled **A–V** on the photograph (**168**) of the base of a skull.

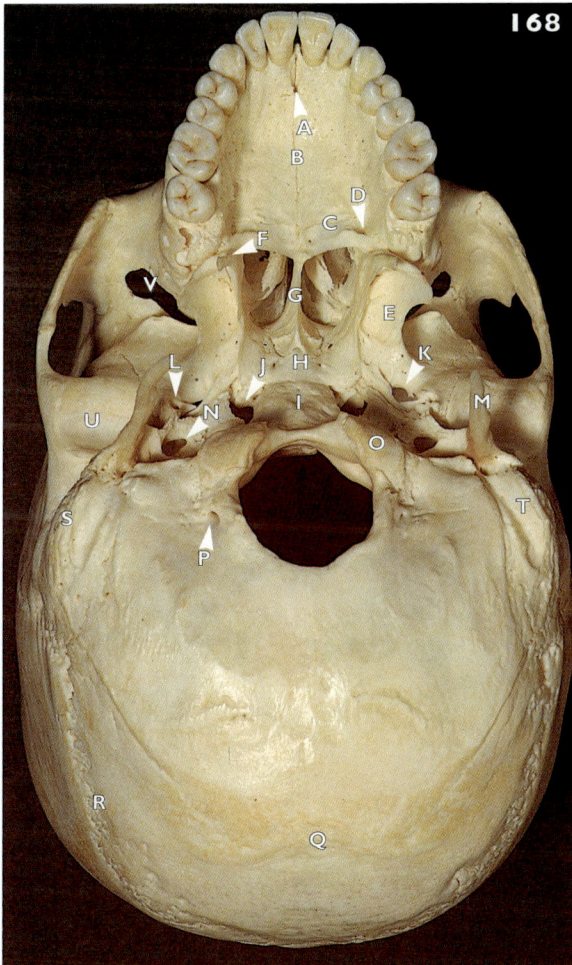

**168** **A:** incisive foramen. **B:** median palatine suture. **C:** transverse palatine suture. **D:** greater palatine foramen. **E:** left lateral pterygoid plate. **F:** right pterygoid hamulus. **G:** vomer. **H:** body of sphenoid bone. **I:** basiocciput. **J:** right foramen lacerum. **K:** left foramen ovale. **L:** right foramen spinosum **M:** left styloid process of temporal bone. **N:** right opening of carotid canal. **O:** left occipital condyle. **P:** posterior condylar canal (for emissary vein). **Q:** external occipital protuberance. **R:** lambdoid suture. **S:** right mastoid suture. **T:** left digastric notch. **U:** right glenoid (mandibular) fossa. **V:** inferior orbital fissure.

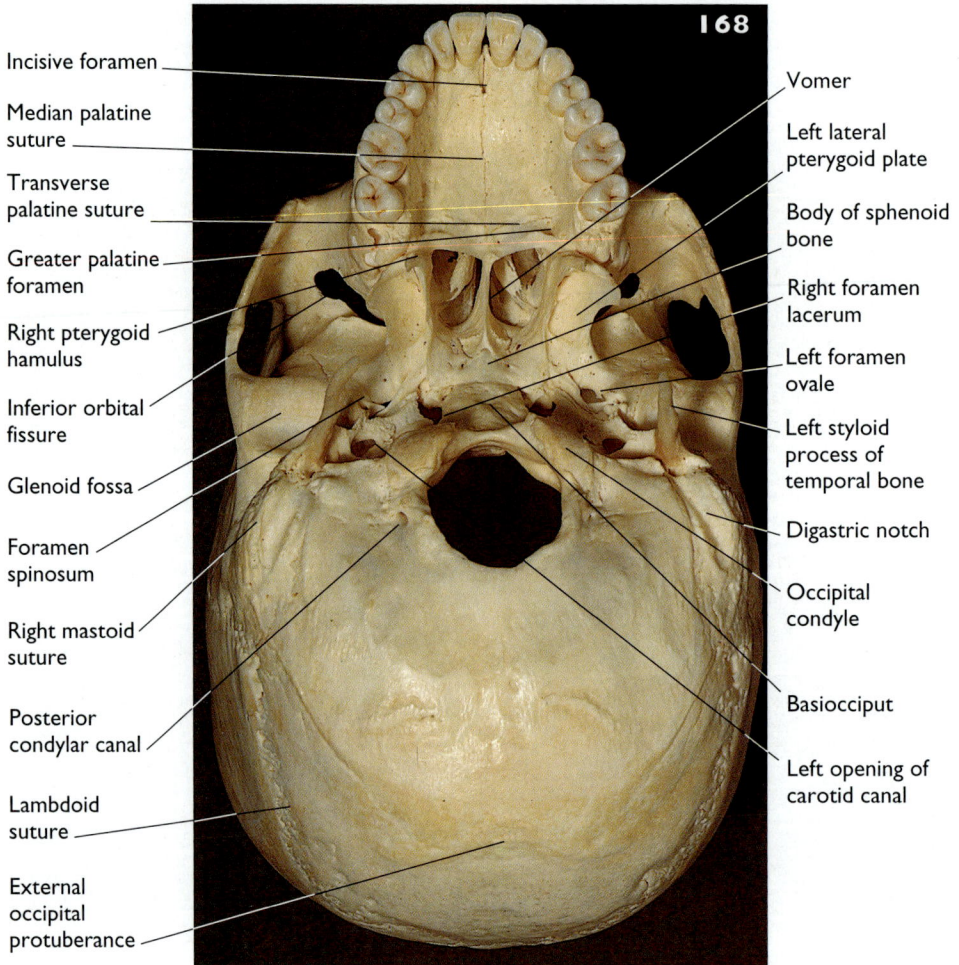

Incisive foramen

Median palatine suture

Transverse palatine suture

Greater palatine foramen

Right pterygoid hamulus

Inferior orbital fissure

Glenoid fossa

Foramen spinosum

Right mastoid suture

Posterior condylar canal

Lambdoid suture

External occipital protuberance

Vomer

Left lateral pterygoid plate

Body of sphenoid bone

Right foramen lacerum

Left foramen ovale

Left styloid process of temporal bone

Digastric notch

Occipital condyle

Basiocciput

Left opening of carotid canal

169  Identify the structures labelled **A–R** on the photograph (**169**) of a superficial dissection of the right side of the face.

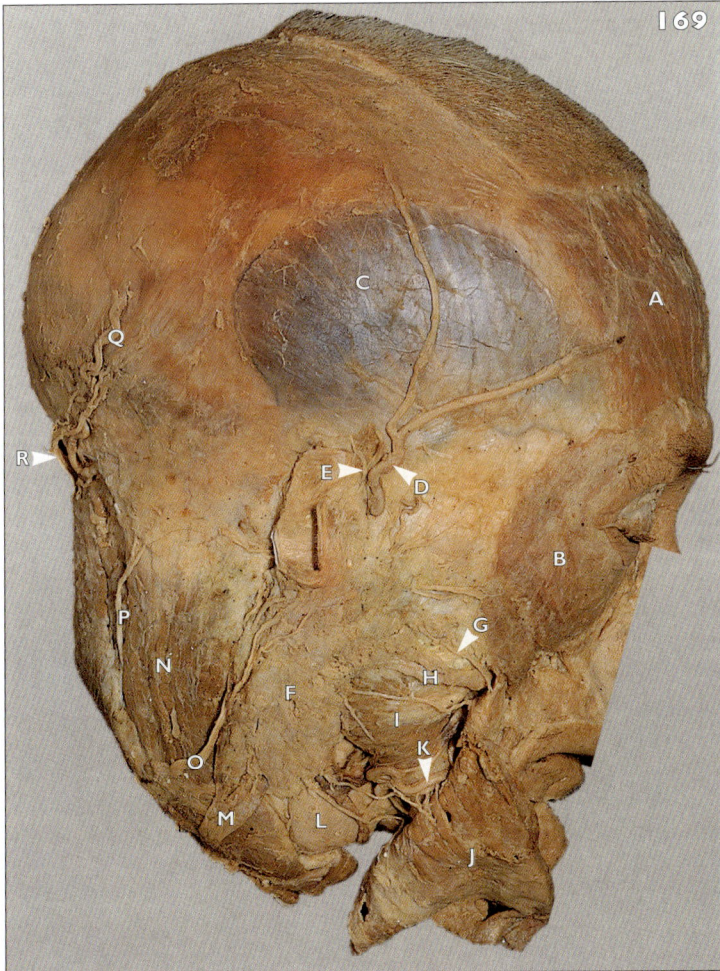

# 169: Answers

**169** A: frontalis. B: orbicularis oculi. C: temporalis. D: superficial temporal artery. E: auriculotemporal branch of mandibular nerve. F: parotid gland. G: a buccal branch of the facial nerve. H: parotid duct. I: masseter. J: reflected platysma. K: facial artery. L: submandibular gland. M: retromandibular vein forming external jugular vein. N: sternocleidomastoid. O: great auricular nerve. P: lesser occipital nerve. Q: occipital artery. R: greater occipital nerve.

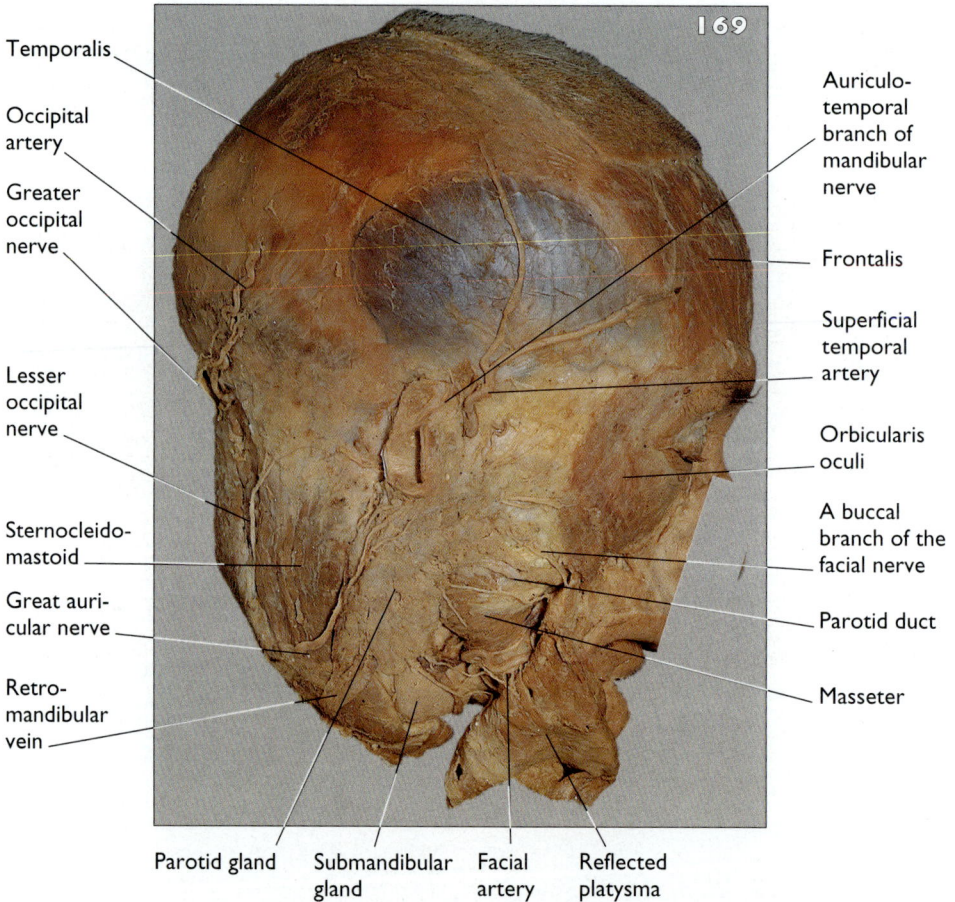

170 Identify the labelled structures (**A–O**) on the dissection (**170**) of the orbital regions.

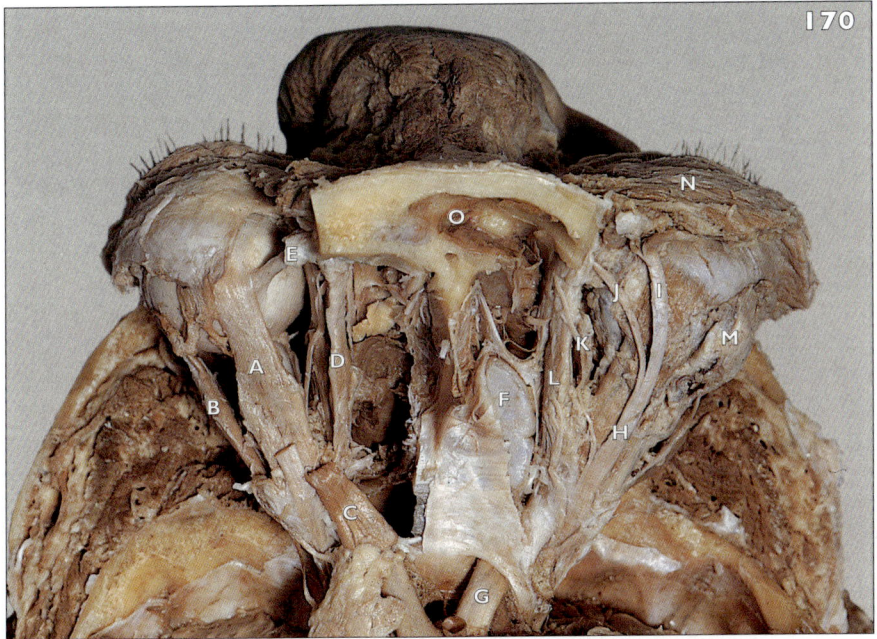

170   A: superior rectus. B: lateral rectus. C: levator palpebrae superioris. D: superior oblique. E: trochlea. F: ethmoidal air sac. G: optic nerve. H: frontal nerve. I: supraorbital nerve. J: supratrochlear nerve. K: nasociliary nerve. L: medial rectus with oculomotor nerve branch. M: lacrimal gland. N: orbicularis oculi. O: frontal air sinus.

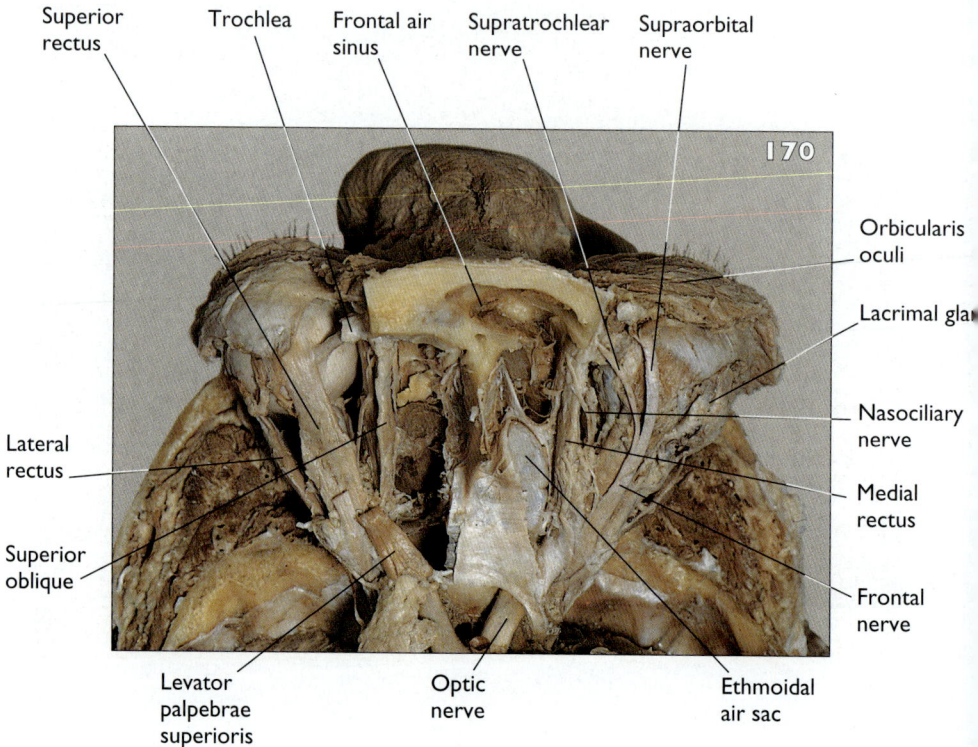

**171** Identify the labelled structures (**A–O**) on the dissection (**171**) of a right infratemporal fossa.

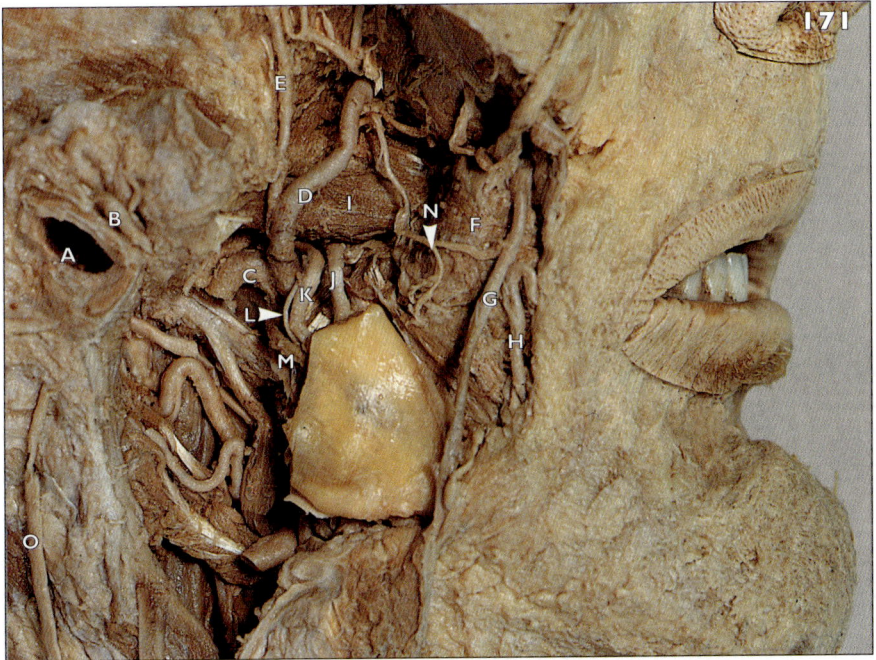

171  A: external acoustic meatus. B: superficial temporal artery. C: maxillary artery. D: middle meningeal artery. E: deep temporal artery. F: buccinator muscle. G: facial vein. H: facial artery. I: inferior head of lateral pterygoid. J: lingual nerve. K: inferior alveolar nerve. L: nerve to mylohyoid. M: inferior alveolar artery. N: buccal nerve. O: great auricular nerve.

Note that there is commonly irregularity with respect to the branches of the maxillary artery in the infratemporal fossa. In this specimen, the maxillary artery appears only briefly in the fossa before travelling deep to the lateral pterygoid. Furthermore, the deep temporal artery is arising from the first part of the artery (i.e. before the lateral pterygoid) and, unusually, before the middle meningeal artery.

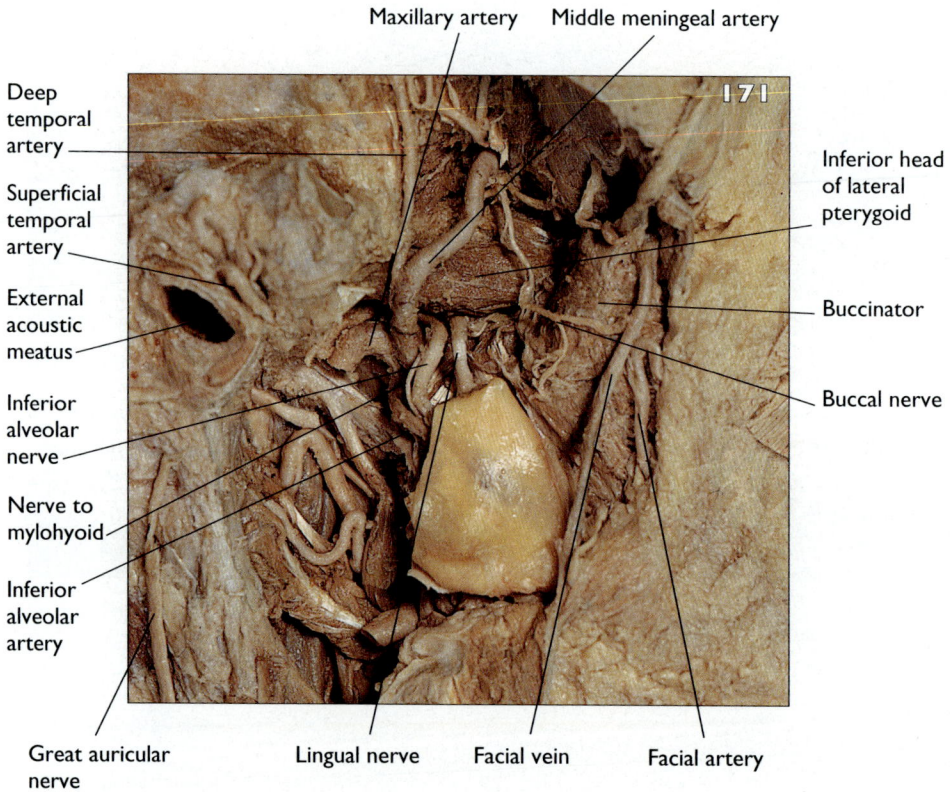

**172** Identify the structures labelled **A–Q** on the coronal head section (**172**).

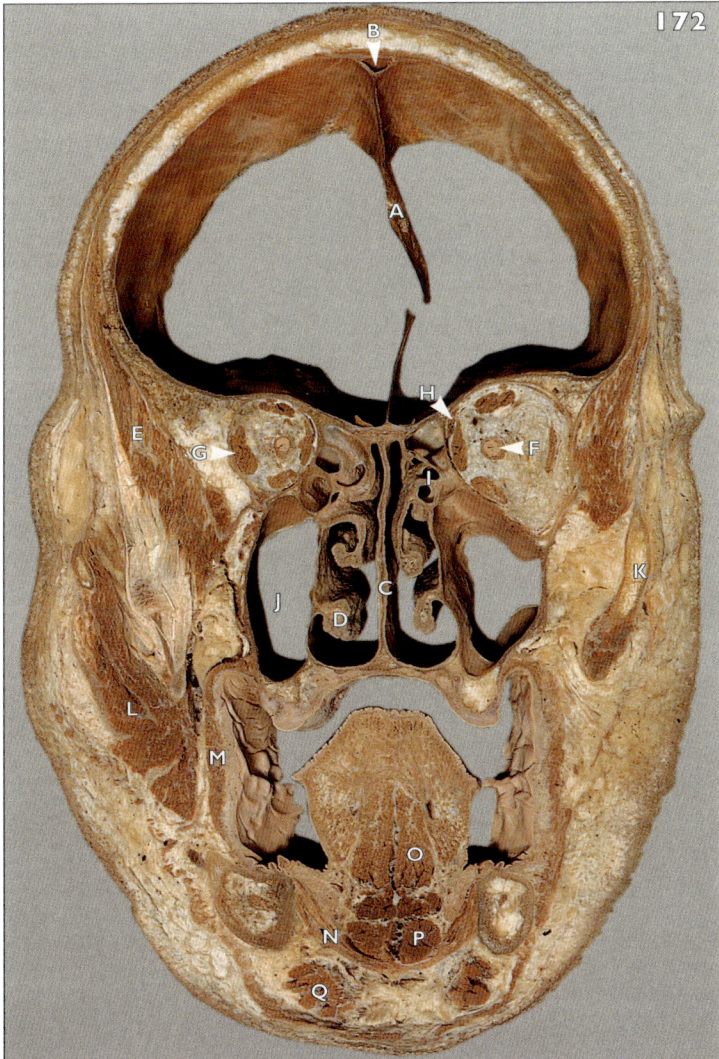

172   A: falx cerebri. B: superior sagittal sinus. C: nasal septum. D: inferior nasal concha. E: temporalis. F: optic nerve. G: lateral rectus. H: superior oblique. I: ethmoidal air cells. J: maxillary air sinus. K: zygomatic arch. L: masseter. M: buccinator. N: mylohyoid. O: genioglossus. P: geniohyoid. Q: digastric (anterior belly).

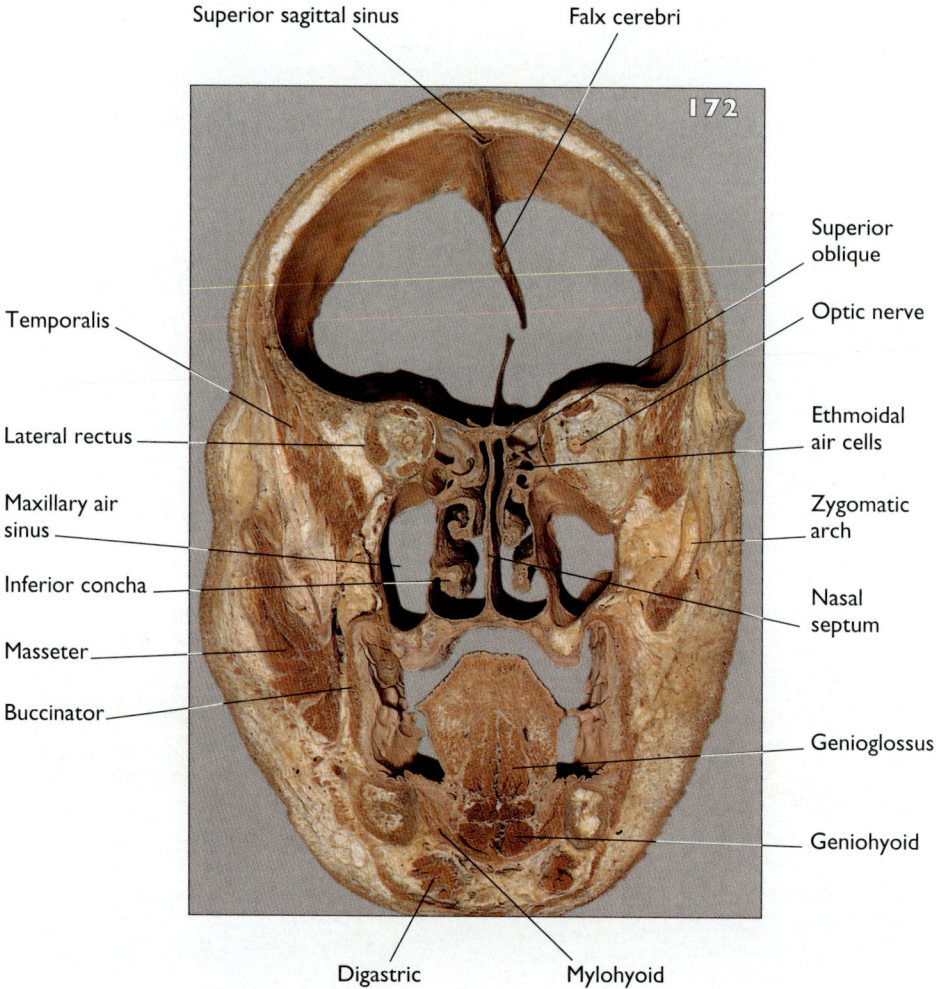

Superior sagittal sinus    Falx cerebri

Superior oblique

Temporalis

Optic nerve

Lateral rectus

Ethmoidal air cells

Maxillary air sinus

Zygomatic arch

Inferior concha

Nasal septum

Masseter

Buccinator

Genioglossus

Geniohyoid

Digastric    Mylohyoid

173 Identify the labelled structures **A–N** on the axial (transverse) section of the head and neck (**173**).

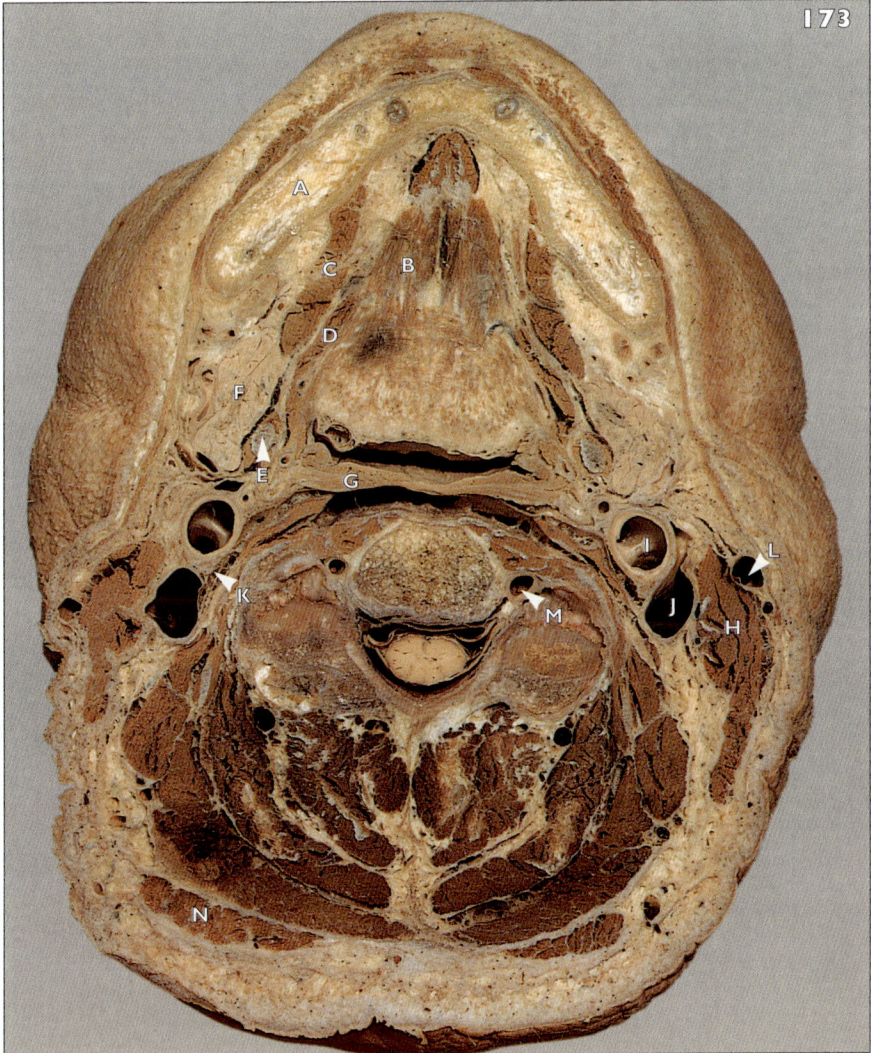

173    A: body of mandible. B: genioglossus. C: mylohyoid. D: hyoglossus. E: inter-mediate tendon of digastric. F: submandibular salivary gland. G: middle constrictor of pharynx. H: sternocleidomastoid. I: common carotid artery (near bifurcation). J: internal jugular vein. K: vagus nerve. L: external jugular vein. M: vertebral artery in transverse process of cervical vertebra. N: trapezius.

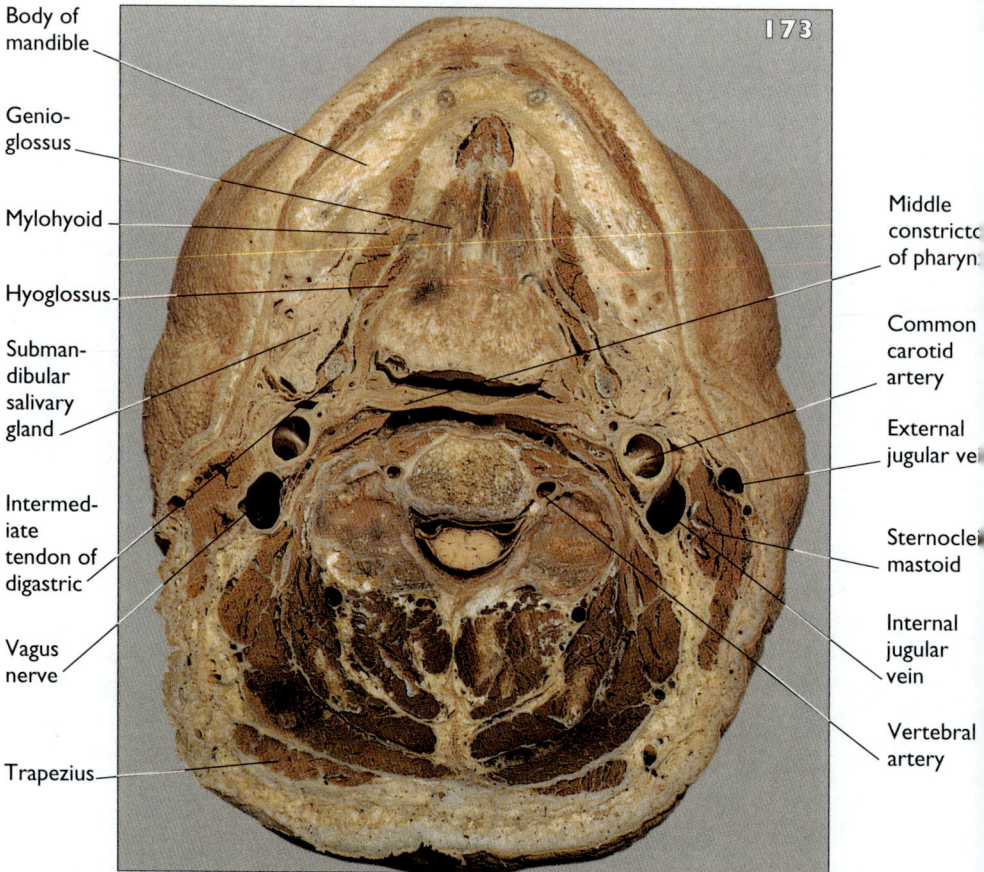

Body of mandible

Genio-glossus

Mylohyoid

Hyoglossus

Subman-dibular salivary gland

Intermed-iate tendon of digastric

Vagus nerve

Trapezius

Middle constrictor of pharynx

Common carotid artery

External jugular vein

Sternocleido-mastoid

Internal jugular vein

Vertebral artery

174 A motorcyclist was involved in a serious road traffic accident. He was rushed to hospital in considerable pain and with suspected fractures of bones in his limbs. He was initially conscious and able to talk to the hospital staff. However, as he was being observed and treated, the patient suddenly became very noticeably drowsy. It was only then realized that there was bruising on his left temple. A radiograph of the skull revealed a fracture running down the squamous part of the temporal bone and across the base of the middle cranial fossa. Re-examination showed that the patient's left pupil was dilated, but not the right pupil, and that there was weakness of his facial musculature. A magnetic resonance image scan was quickly prescribed and this showed that there was blood collecting around the lateral aspect of the left cerebral hemisphere (174).

i. What is happening in this case?

ii. How would the condition be treated in this emergency?

iii. What might be the outcome of delay in treatment?

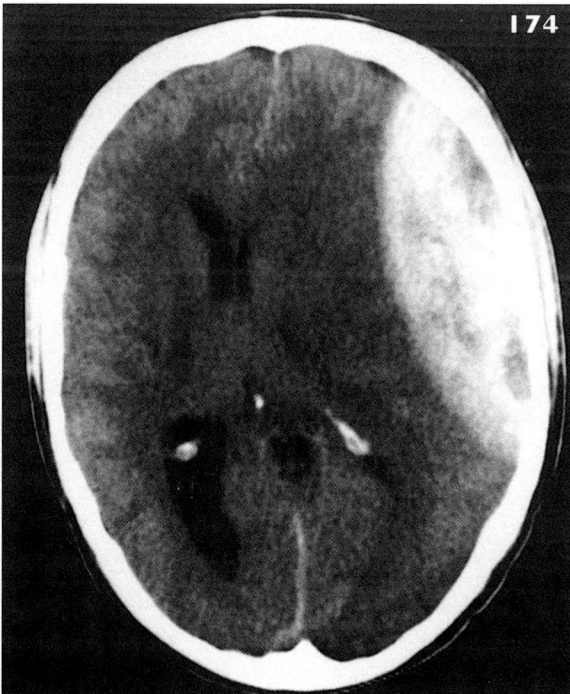

121

174   i. The squamous part of the temporal bone is very thin and prone to fracture. Such a fracture can cause rupture of the left middle meningeal artery within the middle cranial fossa. Consequently, blood escapes between the dura and the bone to produce an extradural haematoma. In this case, the blood also passed outwards (through the fracture) to cause the external bruising. The extradural haematoma reported here compressed the lateral aspect of the brain and the facial palsy could have resulted from the pressure of the blood clot on the primary motor cortex. Alternatively, direct damage to the left facial nerve as it passes through the internal acoustic meatus or the facial canal (of the middle ear) might have occurred, since the line of the fracture was through the floor of the middle cranial fossa. The pupillary dilation results from the oculomotor nerve being pressed against the edge of the tentorium cerebelli, thus disrupting the parasympathetic fibres within it going to the sphincter pupillae.

ii. The patient would be treated by drilling a small hole (a 'burr hole') through the left temporal bone and draining the blood clot. However this procedure alone may not be sufficient to relieve the pressure on the brain tissue and a more extensive craniotomy may be required.

iii. If the pressure is not relieved quickly enough, displacement of the brain stem at the opening of the tentorium cerebelli could force the crus cerebri of the opposite side against the rim of the tentorium cerebelli. This would result in a left-sided paralysis of the body (hemiplegia) as a result of damaging the corticospinal tracts. Further compression of the brain stem produces 'decerebrate rigidity' and fixed dilation of both pupils. Eventually, death could result from vital centres in the medulla oblongata being compressed.

175 A 58-year-old lady was found unconscious at the foot of a flight of stairs, with severe laceration of the back of the head. On admission to hospital, the first radiograph requested was a lateral view of the cervical spine (175a). This showed a displaced flexion fracture of the dens (odontoid process) of the axis (C2), together with possible subluxation at C3/4. The patient soon recovered consciousness, and on repeated examination there was evidence of weakness in all four limbs.

i. Which important neural structure lies immediately posterior to the dens?

ii. How could examination of the ophthalmic division of the trigeminal nerve help test the integrity of this structure?

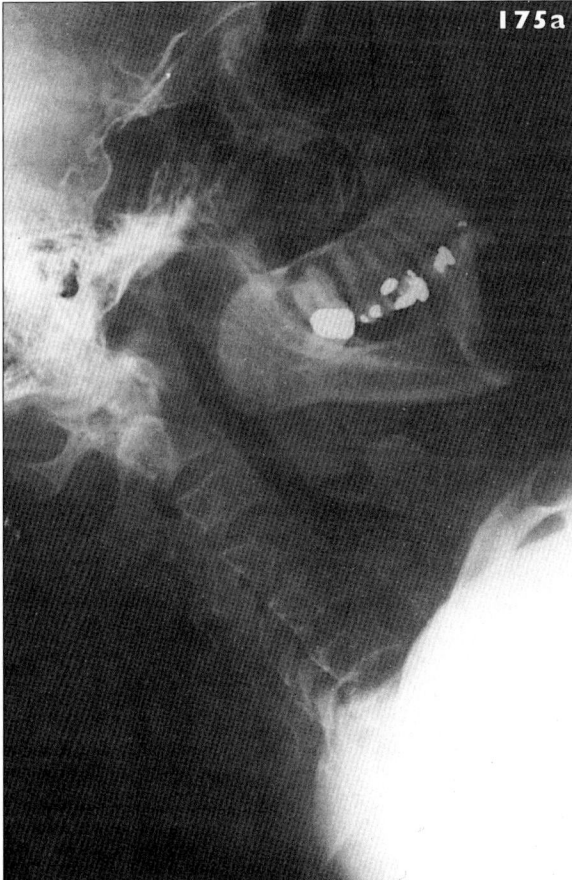

**175** **i.** The medulla oblongata is immediately posterior to the dens. Damage to this structure would be serious because of the presence there of 'vital centres' (e.g. the respiratory centre).

**ii.** The fibres of the ophthalmic division of the trigeminal nerve descend in the lateral part of the medulla oblongata to reach the lower extremities of the nucleus of the spinal tract of the trigeminal nerve. Minor misalignment of the dens may first disrupt these fibres and this will be characterized by loss of pinprick sensation over the cutaneous distribution of the ophthalmic nerve. An accurate sign of this disruption is the corneal reflex. Touching the cornea with a fine wisp of cotton wool normally results in rapid blinking of both eyes, because of initial stimulation of the ophthalmic nerve.

The illustration **175b** shows a sagittal (just off the midline) view of the left half of the head.

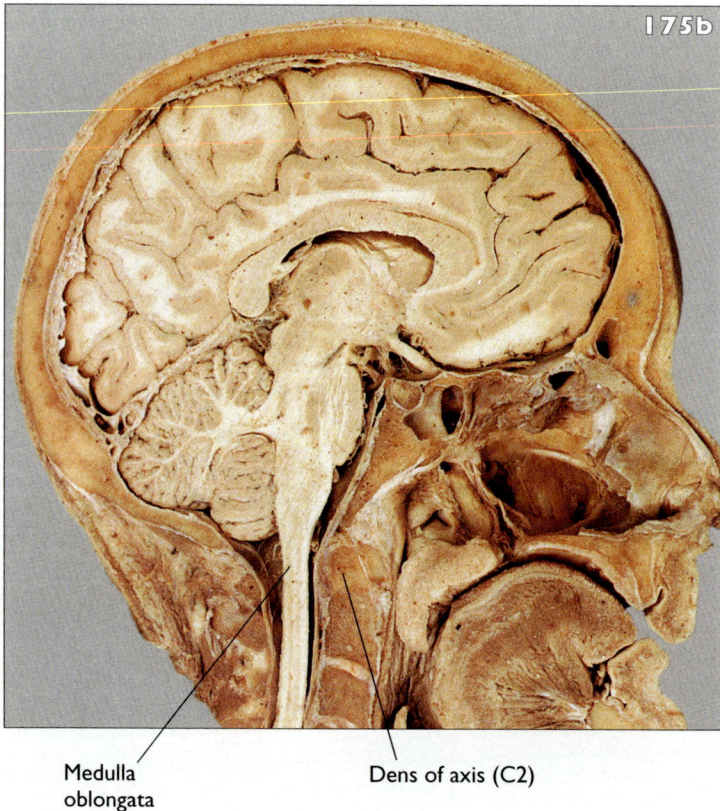

Medulla oblongata          Dens of axis (C2)

**176** A 26-year-old woman was persuaded by her partner to visit the local hospital because of a swelling in her neck which she had been aware of for the previous month. Whilst she was describing the swelling, the doctor noticed that her voice was slightly hoarse, but the patient said that she often had throat infections. On investigation, the doctor observed a non-pulsatile, smooth, soft, well-defined swelling near the midline and in the region of the thyroid gland. He suspected that the swelling was either a thyroid goitre (endogenous enlargement of thyroid tissue) or a thyroglossal cyst close to the thyroid gland. These two conditions can be differentiated in the clinic on the basis of differences in their attachments to surrounding structures in the neck.

i. Why did the doctor ask the woman to stick out her tongue and then to swallow?

ii. Could the patient's hoarseness in fact be connected with the swelling in the neck? How might this connection be investigated?

**177** A doctor examined a young man who, five months previously, had gone over the handlebars of his cycle when braking suddenly to avoid a child who ran out into the road. When the patient had removed his shirt, the doctor saw that his right arm was hanging limply by his side. It was medially rotated, his forearm pronated and there was obvious wasting of the biceps and deltoid muscles. The doctor also discovered that there was a loss of sensation down the lateral side of the patient's arm and hand. When the patient was asked to turn with his back facing the doctor, it was clear that he had a 'winged' scapula on the right side (**177**).

i. What is the extent of the injury that the cyclist has sustained as a result of the accident and how do you explain the defects in the upper limb?

ii. What is the cause of the 'winged' scapula?

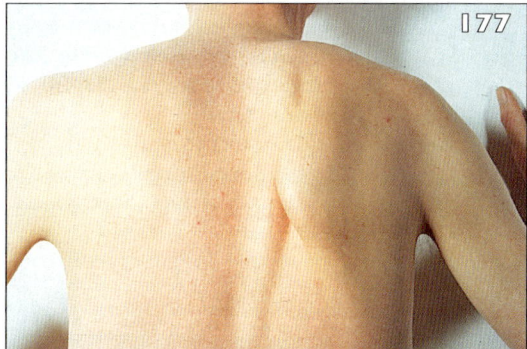

**176** i. Epithelial remnants of the thyroglossal duct from the foetus may occur at any position along the pathway of descent of the developing thyroid gland (**176**). Such remnants may become cystic. Thyroglossal cysts are situated near the midline and are usually attached above to the muscles of the tongue by strands of fibrous tissue. Therefore, when the tongue is stuck out, a thyroglossal cyst can be observed to move upwards. If the swelling is due to a thyroid goitre, the enveloping pretracheal fascia prevents the gland moving upwards on protrusion of the tongue. However, the gland can be seen to move upwards during swallowing since the thyroid gland is bound to the larynx by the pretracheal fascia.

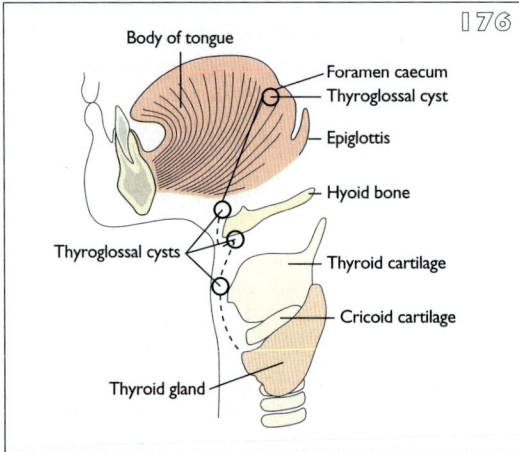

Body of tongue
Foramen caecum
Thyroglossal cyst
Epiglottis
Hyoid bone
Thyroglossal cysts
Thyroid cartilage
Cricoid cartilage
Thyroid gland

ii. If a goitre has grown sufficiently to compress the recurrent laryngeal nerve, the tension of the vocal cords is affected and a hoarse voice is produced. Examination of the vocal cords with a laryngoscope will indicate if the vocal cords are symmetrical and move in synchrony.

**177** i. The brachial plexus has been damaged. Such damage gives rise to the condition termed 'Erb–Duchenne palsy'. When the patient was thrown from his cycle, his head was thrust excessively to the left and his right shoulder was depressed. As a consequence, he tore the roots of the brachial plexus derived from the fifth and sixth cervical nerves. This resulted in paralysis and wasting of several arm muscles supplied by these nerves (i.e. the supraspinatus, infraspinatus, subclavius, biceps brachii, brachialis, coracobrachialis, deltoid, and teres minor). The limb hung down by his side and was medially rotated because of the unopposed action of subscapularis. The forearm was pronated because of the loss of the action of biceps. The loss of sensation down the lateral side of the arm and head is related to the fact that the dermatomes here are associated with the fifth and sixth cervical nerves.

ii. The 'winged' scapula results from paralysis of the rhomboid muscles and serratus anterior (supplied by the dorsal scapular nerve – a branch of the fifth cervical nerve root). These muscles normally retract the scapula.

**178** A medical student was asked to insert an endotracheal tube into a patient under the supervision of a consultant anaesthetist. The anaesthetist explained to the student that, during general anaesthesia, it is often necessary to pass such a tube through the mouth and into the trachea to facilitate ventilation of the patient. The patient, a middle-aged woman who was a heavy smoker, had been sedated and was lying on her back with her head propped up by three pillows. The student, who was performing this procedure for the first time, opened the patient's mouth and inserted a laryngoscope. When he saw the vocal cords just behind the epiglottis, he inserted the endotracheal tube. When the tube was connected to the oxygen supply, the student placed his stethoscope over the chest but was unable to hear any sounds of breathing. At the same time, the consultant noticed a rapidly enlarging swelling in the epigastric region of the patient's abdominal wall.

i. What has gone wrong with the procedure?

ii. Would you expect the patient to experience abdominal pain in such an instance, and what would be the anatomical basis of such pain?

**179** A physician made a home visit to a 59-year-old man. The doctor noticed that the patient had a serious chest infection and, additionally, was in a generally debilitated state. It was clear that the man required hospital admission as soon as a bed could be found. On taking a history in the ward, the doctor discovered that the patient had experienced difficulty in swallowing liquids for some time. The patient said that he had noticed that swallowing became more difficult as the day went on. On examination, the doctor found a swelling on the left side of the man's neck that was soft and fluctuant on palpation. The patient, with some embarrassment, told the doctor that he had recently begun to regurgitate undigested food; he also suffered from halitosis.

i. What is the swelling due to?

ii. Why did the man suffer so badly from bad breath?

iii. Is the chest infection related to the swelling in the neck?

**178 i.** The pharynx is a common chamber leading from the oral and nasal cavities above to the trachea and oesophagus below. The trachea is situated in front of the oesophagus. In the present case, the student had incorrectly placed the endotracheal tube into the oesophagus and the pressure of the anaesthetic gases had caused the stomach to expand. Because the oesophagus lies posteriorly, it is more or less in line with the oral cavity, the trachea being at more of an angle. Therefore, a tube passed blindly down the pharynx is more likely to enter the oesophagus unless precautions are taken. One of the most important of these is to position the inlet of the larynx in line with the oral cavity. This can be achieved by extending the head at the atlanto-occipital joint and flexing the lower part of the cervical spine (**178**). In the case described above, the presence of three pillows had actually flexed the head and neck. In addition, the vocal cords should be clearly visualized before introducing the endo-tracheal tube.

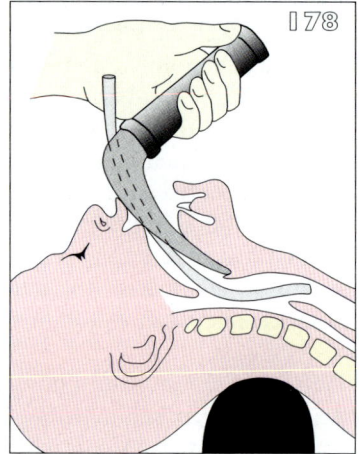

**ii.** The patient would have experienced 'stomach' pain due to the overextension of the stomach wall which would have been detected by afferent receptors in the wall itself and 'referred' to the T8 dermatome.

**179 i.** A weak spot in the posterior pharyngeal wall usually occurs between the two components of the inferior constrictor – i.e. between the transversely arranged fibres of cricopharyngeus and the more oblique fibres of the thyropharyngeus muscle. At this site, the diverticulum is known as 'Killian's dehiscence' (**179**). The diverticulum is free to enlarge laterally, producing a swelling at the side of the neck. It may compress the oesophagus, producing difficulty in swallowing. The opening of the diverticulum can come to lie directly below the pharynx as it descends towards the oesophagus.

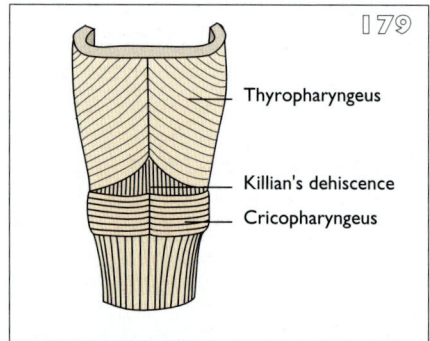

**ii.** Undigested food fills the pouch and can remain there for days before being regurgitated, causing halitosis.

**iii.** When the patient lies down at night, the contents may empty into the oesophagus. Some of the contents may be aspirated into the lungs, producing chest infections.

180 A man visited his doctor a few days after returning from a skiing holiday. The weather during his holiday had been exceptionally cold. He had made an appointment to see his doctor because he had noticed that the left side of his face had become slightly distorted and expressionless towards the end of the holiday. He could not completely close his left eye and was unable to smile properly as that corner of his mouth was drooping. He also had difficulty in retaining food in his mouth while eating. He told the doctor that his daughter had asked him why he wasn't whistling, something he always used to do as the relaxing effects of a holiday began to take effect! On being asked by his doctor to 'screw up his eyes', the facial asymmetry became more obvious (180b).

i. What has caused these symptoms?

ii. Why was the doctor so concerned that the patient could not completely close his eye?

iii. During his investigation, the doctor asked the patient to wrinkle his brow. Why?

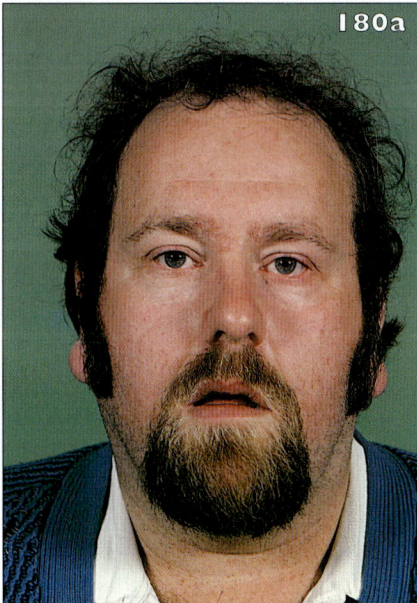

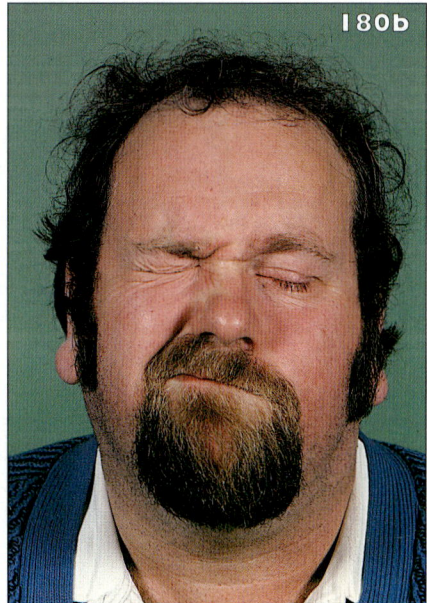

**180  i.** The patient has Bell's palsy, which affects the facial nerve and consequently the muscles of facial expression. In the absence of obvious trauma, the aetiology of this condition is unknown, although it is commonly associated with being in a cold or draughty environment. A hemiparalysis of the muscles of facial expression leads to a loss of the normal creases and folds in the face on the affected side, giving the face an almost expressionless (death mask) appearance. Paralysis of the muscles associated with the cheek and lips was responsible for the inability to smile, whistle, and eat properly.
**ii.** The loss of action of the orbicularis oculi muscle and the absence of blinking can render the conjunctiva and cornea susceptible to inflammation and may eventually lead to blindness, if untreated. It is therefore advisable to protect the 'open' eye when the patient goes to sleep at night.
**iii.** A facial palsy can also result from an upper motor neuron lesion. Where only the lower half of the face is affected, a lesion in the contralateral side of the brain above the facial nucleus in the pons is indicated, as the upper half of the face has some ipsilateral as well as contralateral representation.

Bell's palsy can also be caused by a viral infection of the facial nerve which may be extensive. In addition to the hemiparalysis of the facial muscles the patient may suffer some reduction in the sense of taste and in salivation as a result of interruption of the afferent and parasympathetic fibres respectively, which run in the facial nerve. The patient may also experience abnormally loud sounds (hyperacusis) on the affected side resulting from paralysis of the stapedius muscle. Lacrimation, caused by other parasympathetic fibres in the facial nerve branch which passes to the pterygopalatine ganglion, is rarely associated with a Bell's palsy.

**181** A young man visited his dentist complaining of a painful swelling of the soft tissues around an erupting 'wisdom' tooth in his lower jaw (**181**). An infection arising from a flap of soft tissue surrounding the crown of the erupting tooth (pericoronitis) was diagnosed, and the region was syringed out. A week later, the patient returned with the same problem and the treatment was repeated. Two weeks later, he was admitted to hospital in considerable distress and, despite emergency treatment, died shortly afterwards. The patient had had

trismus (difficulty in opening his jaw), difficulty in swallowing and breathing, and a high temperature (40°C). There was also considerable swelling of the soft tissues in the face and neck, and at the back of the mouth. What had happened and how do you account for the signs and symptoms?

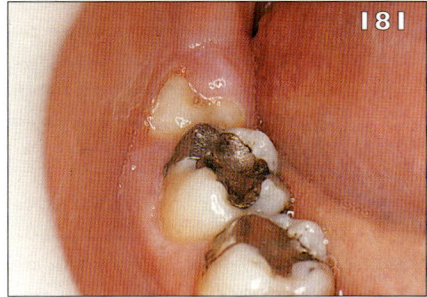

**182** One morning while shaving, a 65-year-old steelworker noticed in the mirror that he was unable to elevate his left upper eyelid. In fact, the eyelid drooped (ptosis) (**182** illustrates this condition in a young child). As he was concerned about this, he made an appointment to see his doctor. In the surgery, he also told the doctor that the left side of his face was flushed and felt warm, but that he did not sweat noticeably on the affected side (anhidrosis), even in the intense heat of the steelworks. The doctor observed that the pupil of the man's left eye was constricted (miosis) and did not dilate when he shone a torch into the eye; the other eye was unaffected. It was reported in the man's medical record that he was a very heavy smoker. He admitted that he still smoked

heavily and told the doctor that he had had a persistent cough over the past few years. The doctor referred the patient for a chest radiograph, and a shadow was discovered at the apex of the left lung, indicative of a cancer. How might the information from the radiograph explain the patient's facial symptoms?

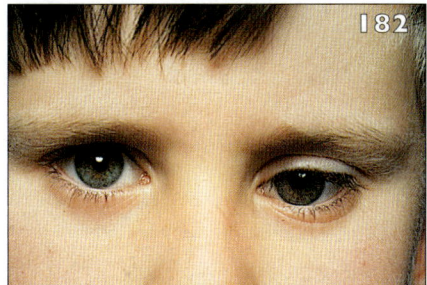

# 181 & 182: Answers

181 The infection spread from the tooth to the tissue spaces associated with the infratemporal fossa, i.e. the pterygomandibular space and the superior part of the pharyngeal tissue spaces. The pterygomandibular space is located between medial pterygoid and the ramus of the mandible. Infection in this site is restricted inferiorly by the attachment of the muscle into the angle of the mandible; thus, infection is prevented from spreading from the pterygomandibular space directly into the neck. On the other hand, the parapharyngeal space between the medial pterygoid and the superior constrictor of the pharynx will allow inflammatory products to spread directly into the neck. In the case described here, the trismus indicates that the infection had spread into the infratemporal fossa. The infection had then passed into the neck, as evidenced by the difficulty in swallowing and breathing (swelling impinging on the airway and oesophagus). The swelling at the back of the mouth suggests that the infection had also spread between the superior constrictor and the mucosa of the mouth into the peritonsillar region. Death was probably caused by asphyxia.

182 The facial symptoms indicate that there has been involvement of the left sympathetic chain. The sympathetic outflow to the head and neck has its origin in the thoracic region. The thoracic sympathetic trunk lies on the necks of the ribs and continues upwards over the neck of the first rib as the cervical sympathetic chain. The radiograph revealed that the patient had an apical carcinoma of the left lung. The carcinoma had spread locally to involve the sympathetic trunk at the thoracic inlet, interrupting the preganglionic sympathetic efferents before they reach the superior cervical ganglion. The levator palpebrae superioris muscle and the dilatator pupillae muscle have a sympathetic innervation. The sweat glands on the face are also innervated by the sympathetic system. Interference with the sympathetic innervation of blood vessels results in their vasodilation, making the face red and warm. Hence, all the facial signs can be explained in terms of sympathetic dysfunction. The ptosis, miosis, and anhidrosis caused by paralysis of the cervical sympathetic system is known as Horner's syndrome. An additional sign associated with this syndrome is enophthalmos – a recession of the eyeball into the orbit. When the tumour also affects the brachial plexus, the syndrome is called Pancoast's syndrome.

**183** A young woman went to her local dentist complaining of a generalized, dull toothache in the left maxillary region and of very sharp pains in her forehead. After examining her teeth, the dentist proclaimed her to be fully dentally fit. However, he ascertained that there was a feeling of 'fullness' in her left cheek and so he suspected that she was suffering from maxillary sinusitis and, in relation to the pains in her forehead, acute frontal sinusitis. He referred her to an ear, nose, and throat specialist who confirmed the diagnosis from a radiograph (**183a**) and from the observation of pus in the left middle meatus of the nose.

i. The radiograph usually taken to examine the air sinuses is an occipitomental (OM) projection. Why would a standard anteroposterior view of the skull (AP skull) be inappropriate for this type of examination?

ii. What feature on the radiograph illustrated is 'diagnostic' of acute frontal sinusitis?

iii. How would a physician examine the inferior, middle, and superior meatus of the lateral wall of the nose?

iv. Why would pain from the maxillary sinus be confused by the patient as toothache?

v. Where is the opening of the maxillary air sinus (ostium) and why is its position disadvantageous for drainage?

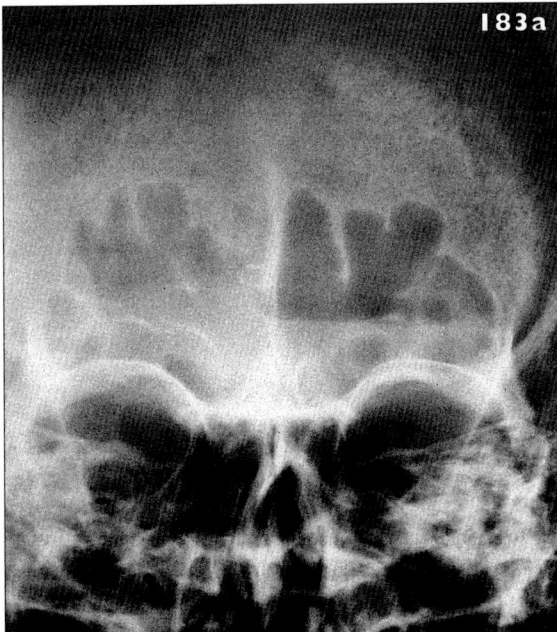

183a

**183** i. An AP radiographic view of the skull will show the sinuses but there will be much superimposition of structures, particularly for the maxillary, ethmoidal, and sphenoidal sinuses. The OM view, by appearing to 'tilt' the head upwards, significantly reduces superimposition of the sinuses.

ii. Note that there appears to be a 'fluid level' within the frontal sinus. This is confirmed by tipping the head to one side and taking a further radiograph (**183b**).

iii. The inferior and middle meatuses are visible by simply viewing the nasal cavity through the nostrils after dilatation with a speculum. This procedure is known as 'anterior rhinoscopy'. However, to see the superior meatus (as well as the inferior and middle meatuses) it is necessary to place a small mirror at the back of the mouth and, in order to bring the soft palate forward, to ask the patient to breathe through the nose.

iv. The superior alveolar nerves and the infraorbital nerve are all branches of the maxillary division of the trigeminal nerve that innervate both the maxillary air sinus and the maxillary dentition.

v. The ostium of the maxillary air sinus is located on its nasal (medial) wall, close to its superior 'margin'. Thus, because of the normal, upright posture of the head, the position of the ostium is unfavourable for drainage which cannot rely upon gravity. Drainage therefore requires the effective functioning of the cilia in the respiratory mucosa of the air sinus; drainage is effected by passing mucosal secretions up to the ostium.

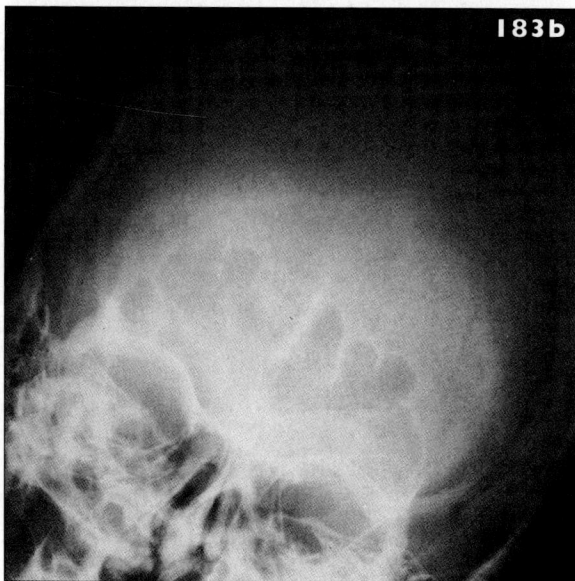

183b

184 At mealtimes, a 45-year-old woman noticed that the right side of her mouth became swollen. She further noticed that the region below her tongue (the floor of the mouth) was painful and that she had a bad taste in her mouth. Her physician examined her mouth and observed that there was inflammation and swelling around the sublingual papilla. A stone or 'sialolith' in a salivary duct was diagnosed (184a) and this was confirmed by radiography (184b).

i. Which salivary duct is involved, and what is its course in the floor of the mouth?

ii. How do you account for the symptoms experienced?

iii. What are the sites of opening of all the major salivary glands?

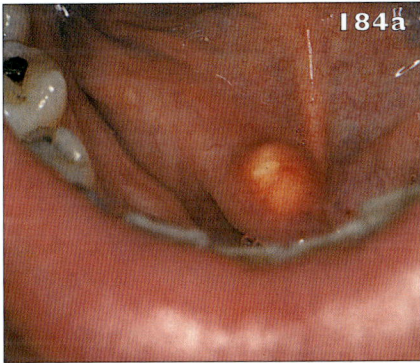

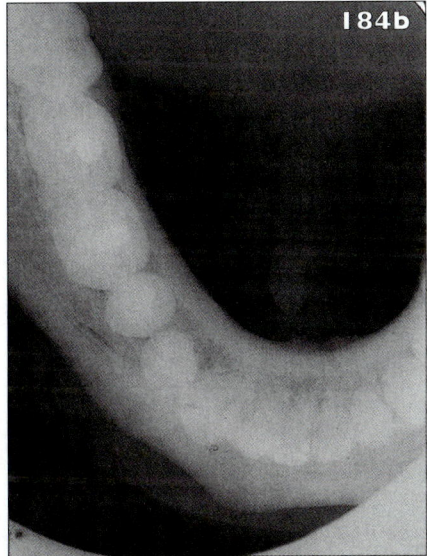

185 When using an auroscope, what would you expect to see in the region of the tympanic membrane?

**184 i.** The submandibular salivary duct is involved, close to its opening on the sub-lingual papilla. The submandibular duct (Wharton's duct) arises in the superficial part of the submandibular salivary gland but emerges from the deep part of the gland as it lies on hyoglossus. The duct crosses the superficial surface of this muscle where, en route, it is crossed twice by the lingual branch of the mandibular division of the trigeminal nerve. The terminal part of the duct lies just beneath the oral mucosa on the sublingual fold.

**ii.** The stone obstructs the flow of saliva and therefore causes the gland to swell as a result of increased secretion at mealtimes. The bad taste is the result of infection.

**iii.** Apart from the submandibular gland (whose duct opens at the sublingual papilla in the floor of the mouth), the major salivary glands include the parotid and sublingual glands. The duct of the parotid gland opens into the cheek (opposite the maxillary second molar tooth). The sublingual gland can be subdivided into two parts, posterior and anterior: the minor ducts of the anterior part form a main duct, which either enters the submandibular duct or appears on the sublingual papilla; the posterior part has many small ducts, which enter the floor of the mouth directly at the sublingual folds.

**185** The illustration (**185**) shows the extent to which the tympanic membrane can be visualized using an auroscope. The membrane appears as a pearly white concave disc. A reddish-brown streak in the depth of the concavity marks the position of the handle of the malleus ear ossicle; this feature is termed the umbo. On illumination, a 'cone of light' is seen in the antero-inferior part of the membrane. A white spot above the umbo indicates the site of the lateral process of the malleus. Furthermore, the long process of the incus ear ossicle may appear as a whitish streak posterior to the upper part of the umbo. Also discernible are the anterior and posterior malleolar folds and the pars flaccida.

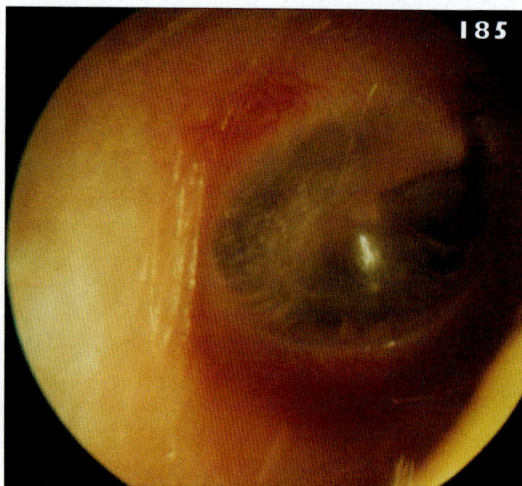

**186** Describe the cutaneous nerve fields on the face associated with the following (**186**):
**i.** The supratrochlear nerve.
**ii.** The infraorbital nerve.
**iii.** The auriculotemporal nerve.
**iv.** The mental nerve.
Name **A–J**. Why is a knowledge of the cutaneous innervation of the face important clinically?

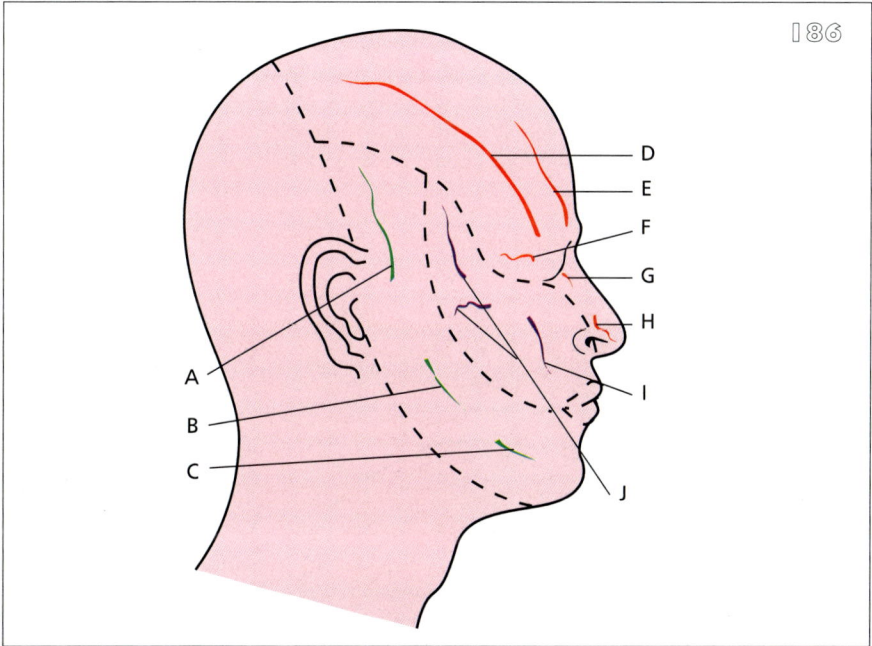

**186 i.** The supratrochlear nerve is a terminal branch of the frontal nerve of the ophthalmic division of the trigeminal. It has a small cutaneous nerve field over the medial part of the forehead and the medial part of the upper eyelid.

**ii.** The infraorbital nerve is the terminal branch of the maxillary division of the trigeminal nerve. It is the largest cutaneous branch of the maxillary nerve. It supplies skin overlying the body of the maxilla, the lower eyelid, the ala of the nose, and the upper lip.

**iii.** The auriculotemporal nerve is a branch of the mandibular division of the trigeminal nerve. It supplies the skin overlying the tragus, concha, external acoustic meatus, and tympanic membrane of the ear. It also supplies the 'beard' (posterior) part of the temporal region.

**iv.** The mental nerve is also a branch of the mandibular division of the trigeminal nerve. It is the terminal branch of the inferior alveolar nerve. It supplies the skin overlying the lower lip and the mandible. Remember, however, that the skin overlying the angle of the mandible is supplied by the great auricular nerve from the cervical plexus.

Knowledge of the cutaneous innervation of the face is important clinically in order to understand the effects of nerve damage and of local anaesthesia (particularly for dental treatment), and to appreciate the tracts taken by vesicles following herpes zoster infections (i.e. shingles).

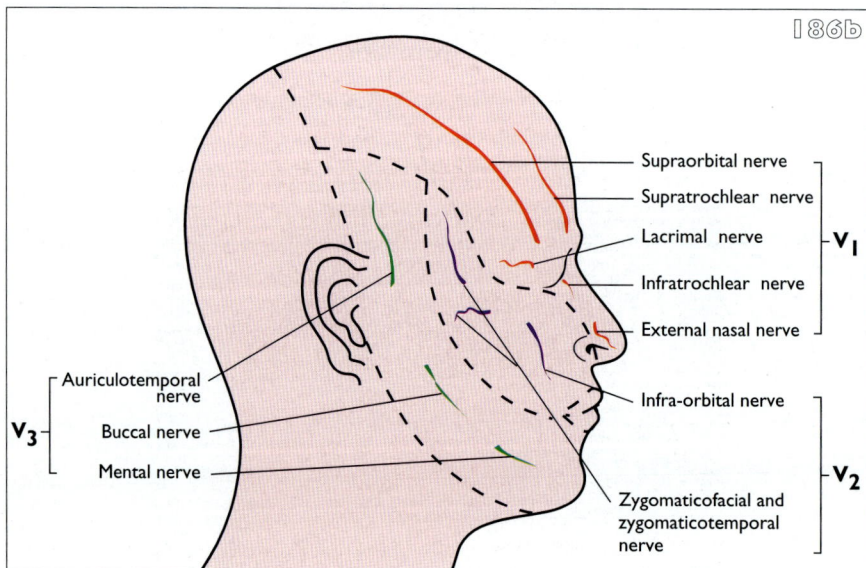

186b

138

**187** A 45-year-old university lecturer in history made an appointment with his doctor as he was very worried that he was becoming unable to do his job satisfactorily. He told the doctor that he had been finding it difficult to express himself properly in lectures and also in comprehending what his students were saying to him in tutorials. When researching his latest book, he was aware that his handwriting was deteriorating and that he had difficulty in using a ruler for underlining. He also told the doctor that his eyesight had worsened. The doctor arranged for the lecturer to have radiographic investigation as a matter of urgency. The axial magnetic resonance image (**187**) revealed a tumour of the left occipital bone with metastases principally in the parieto-occipital regions of the cerebral hemispheres of that side.

**i.** Why was the patient having difficulty in communicating with his students?

**ii.** Why did he find it difficult to use a ruler and why was his handwriting deteriorating?

**iii.** Why was his eyesight failing?

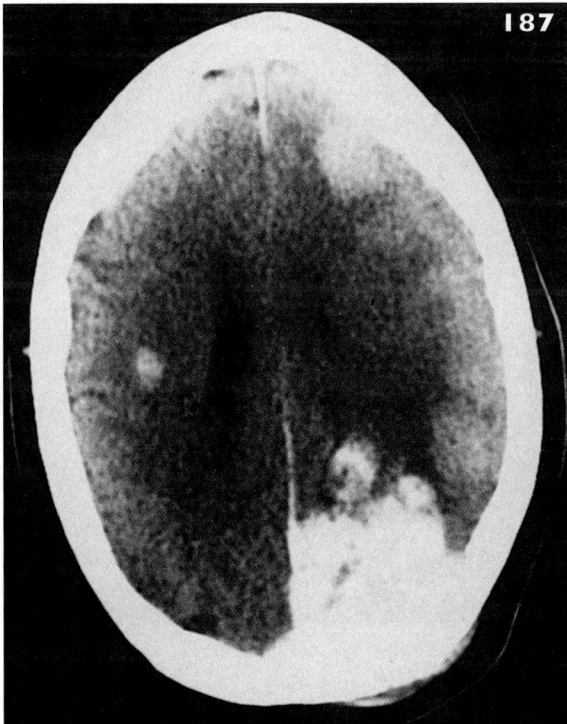

**187  i.** Metastases have affected Wernicke's area in the inferior part of the parietal and posterosuperior part of the temporal lobes of the brain. This is the area on the left side that is responsible for interpreting language by both auditory and visual stimuli; hence the difficulties in lectures and tutorials.

**ii.** The incoordination of his upper limb resulted from damage to the primary motor cortex.

**iii.** His failing sight was due to destruction of the occipital cortex by the ingrowing cancerous bone. The primary visual cortex is situated on either side of the calcarine sulcus in the occipital lobe and is compressed by the growth, thereby cutting off the blood supply from branches of the posterior cerebral artery, which causes the death of the visual cortical tissue.

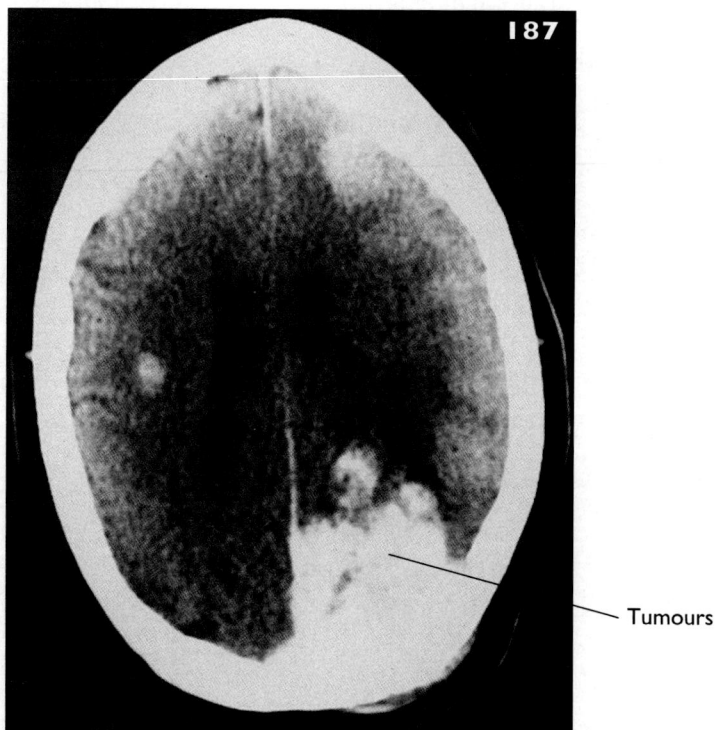

Tumours

188 A patient attends a clinic within the dental hospital complaining of a clicking sound in his jaw joint. The specialist clinician could discern no unusual morphology on palpation nor unusual functioning of the temporomandibular joint (TMJ). Furthermore, the patient did not complain of pain. The patient was sent to the radiography department for a transcranial temporomandibular radiograph (188a and 188b).

i. What are the 'special' features of the TMJ?

ii. Is the pattern of movement shown in the radiographs unusual?

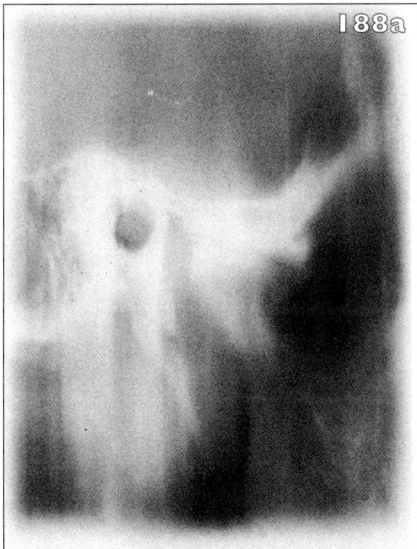

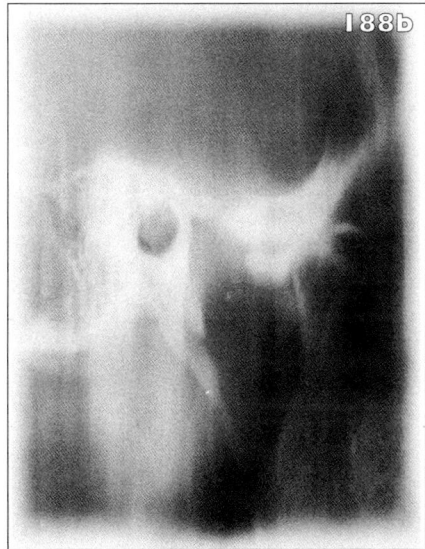

**188   i.** The TMJ is a synovial joint but, because it develops in membrane, the articular surfaces are lined by fibrous tissue and not by hyaline cartilage. During development, the tendon of the lateral pterygoid is 'trapped' within the joint cavity and this gives rise to the intra-articular disc, which is said to subdivide the joint into an upper, gliding compartment and a lower, hinge compartment. A third feature relates to the fact that, although there are accessory ligaments for the TMJ, it is thought that none has a significant influence upon mandibular movements. Finally, because the mandible is a single bone, movements in the TMJ on one side produce similar or 'compensatory' movements in the joint on the other side.

**ii.** The pattern of movement shown in the radiographs (**188a** and **188b**) is normal. Note that, on full opening, the condyle of the mandible not only shows a hinge movement but also glides down the (anterior) articular eminence. In some malfunctioning conditions, only a hinge movement might occur. Movement of the condyle beyond the articular eminence leads to dislocation of the jaw.

# Index

# Index